PRACTICAL ASPECTS OF HYALURONAN BASED MEDICAL PRODUCTS

PRACTICAL ASPECTS OF HYALURONAN BASED MEDICAL PRODUCTS

J. W. Kuo

Boston, Massachusetts, U.S.A.

Taylor & Francis

Taylor & Francis Group

Boca Raton   London   New York

A CRC title, part of the Taylor & Francis imprint, a member of the
Taylor & Francis Group, the academic division of T&F Informa plc.

Published in 2006 by
CRC Press
Taylor & Francis Group
6000 Broken Sound Parkway NW, Suite 300
Boca Raton, FL 33487-2742

Library of Congress Cataloging-in-Publication Data

Kuo, J. W. (Jing-wen)
    Practical aspects of hyaluronan based medical products / J.W. Kuo.
        p. ; cm.
    Includes bibliographical references.
    ISBN-13: 978-0-8493-3324-8
    ISBN-10: 0-8493-3324-5
    1. Hyaluronic acid--Therapeutic use. [DNLM: 1. Hyaluronic Acid--chemistry. 2. Hyaluronic Acid--therapeutic use. 3. Biocompatible Materials. QU 83 K95p 2005] I. Title.

RM666.H66K86 2005
615.5'3--dc22

2005011379

Taylor & Francis Group
is the Academic Division of T&F Informa plc.

Visit the Taylor & Francis Web site at
http://www.taylorandfrancis.com

and the CRC Press Web site at
http://www.crcpress.com

To my wife Lynn

for her love, patience and support

Preface

Natural law is intangible and elusive, and yet within are image and form.

Daode Jin, Chapter 21, Laozi, 6th century B.C.

Hyaluronan (hyaluronic acid, HA) is a giant molecule that has generated huge interest. Its chemical and physical properties as well as its biological behavior have been widely investigated for many years and yet still remain intriguing. Extensive research and development efforts dating from its discovery over seven decades ago have led to the marketing of several HA-based medical products with varying degrees of commercial success. As the database of HA products expands, it has become more feasible to connect the dots of the basic science of HA and the design and performance of HA-based products, which is precisely the intention and approach of this book.

Practical Aspects of Hyaluronan Based Medical Products focuses on the basic properties of HA and on the major characteristics of HA-based products and how their usefulness in medical applications is evaluated in the real world. I hope that readers coming into this field with different backgrounds will find the book helpful for providing the needed conceptual and factual knowledge of the subject and readers more knowledgeable in HA-related research, development, and medical applications will come across useful information and new perspectives as well.

The clinical data and experience of HA-based medical products have such significance that they are the ultimate validation of the original product design or, as the case may be, invalidation. This book focuses on four types of HA-based products as a surgical aid and tissue protective agent in ophthalmic surgery, an injectable preparation to alleviate pain of knee osteoarthritis, a barrier in solid or solution form to prevent postsurgical adhesions, and an injectable gel for cosmetic tissue augmentation. These types of products have been extensively reviewed before their regulatory approval in the United States, and their postmarketing clinical information is being continually generated and made available. There are new areas of HA-based product application under development, such as the use of HA in drug delivery and tissue engineering, which have shown promising prospects. It is hoped that a systematic examination of the basic science of HA and the characteristics and the development history of the existing HA products will benefit the future efforts of HA-related product development.

I wish to take this opportunity to express my sincere acknowledgment to Dr. David Swann, a renowned HA researcher and entrepreneur, who kindly introduced me to this interesting area of research and development, guided me with his profound vision and knowledge, and challenged me with ideas and responsibilities, as well as to Dr. Glenn Prestwich, a prolific chemistry professor and my academic advisor, who provided invaluable direction and coaching in my Ph.D. research of the chemical modification of HA, and who continues to inspire me with his new and innovative research in HA chemistry. Special thanks also to Dr. Bryan Toole, Dr. Kenneth Brandt, and Dr. Arthur Haney, with whom I had the privilege to collaborate on HA-related projects and benefited tremendously from their knowledge and wisdom. The timely completion of this book also is owed to the technical assistance of Dr. Li Jing Sun and Sharon Kuo, which I deeply appreciate.

Finally, I am extremely gratified to have the manuscript completed at the time of the centennial anniversary of my father, P. K. Kuo, M.D., who devoted his entire adult life to the advancement of ophthalmology.

The Author

J. W. Kuo, Ph.D. is an independent researcher and consultant, who has focused on HA-related science, technology, and product development. He is a member of the American Chemical Society.

For his doctoral degree, Dr. Kuo conducted HA-related research under the guidance of Dr. Glenn D. Prestwich in the department of chemistry, State University of New York (SUNY) at Stony Brook. He worked as director of basic research (1989–1992), vice president of research and development for MedChem Products, Woburn, Massachusetts (1992–1993), and vice president of research and development for Anika Therapeutics, also in Woburn (1993–1995), under the direction of the founder and CEO of both companies, Dr. David A. Swann. Dr. Kuo continued his service with Anika Therapeutics as its vice president of technical and clinical affairs (1995–1998). In working with Spray Venture Capital in Boston, he was active in the start up of Exhale Therapeutics, an HA company, as a consultant (1999–2000), and he served as its vice president of chemistry from 2001 to 2002.

The author has over 20 years of industry experience in the development of HA-related products and technology, including: (1) R&D effort in the chemistry, chemical modification, formulation, and testing of HA based products for new applications; (2) manufacturing technology involving the isolation, purification, and characterization of HA from various sources; (3) collaboration with academic and medical institutions in activities relating to preclinical and clinical studies; (4) the safety and performance assessments pertinent to such products, and the draft of their technical reviews for submission to regulatory agencies in the U.S. and EU; and (5) the quality assurance and quality control of HA operations in compliance with FDA Quality System (QS) regulation and ISO standards.

Dr. Kuo has authored and coauthored a number of research papers in peer-reviewed journals, most of which involve hyaluronic acid, and he has been issued, as a co-inventor, several hyaluronic acid–related patents by the United States Patent and Trademark Office.

List of Abbreviations

ACLT	anterior cruciate ligament transection
AE	adverse events
AFS	American Fertility Society
BDDE	1,4-butanediol diglycidyl ether
BODIPY	4,4-difluoro-4-bora-3a,4a-diaza-s-indacene
BSS	balanced salt solution
Cbz	carbonylbenzyloxy
CD	circular dichroism
CDI	carbodiimides
CDRH	Center for Devices and Radiological Health
cft	capillary flow times
CMC	carboxymethylcellulose
CNBr	cyanogen bromide
cP	centiPoise
CP-MAS	cross polarization magic angle spinning
CS	chondroitin sulfate
cSt	centiStoke
CV	coefficient of variations
Da	dalton (unit of molecular weight)
DEAE	diethylaminoethyl
DMAB	dimethylaminobenzaldehyde-Ehrlich reagent
DMSO	dimethyl sulfoxide
DTP	3,3'- dithiobis(propionic hydrazide)
DVS	divinylsulfone
ECM	extracellular matrix
EDC	1-[3-(dimethylamino)propyl]-3-ethylcarbodiimide
EDDE	1,2-ethanediol diglycidyl ether
EDU	1-[3-(dimethylamino)propyl]-3-ethylurea
ELISA	enzyme-linked immunosorbent assay
EP	European Pharmacopoeia
ESI-MS	electrospray-ionization mass spectrometry
ESR	electron spin resonance
EU	endotoxin unit

EULAR	European League of Associations of Rheumatologists
FA	5-aminofluorescein
FACE	fluorophore-assisted carbohydrate electrophoresis
FDA	Food and Drug Administration
FTIR	Fourier Transform infrared spectroscopy
GAG	glycosaminoglycan
GLC	gas liquid chromatography
GlcNAc	N-acetylglucosamine
GlcUA	glucuronic acid
HA	hyaluronic acid, hyaluronate or hyaluronan
HABP	HA binding protein
HAS	HA synthase
HDA	hexyldiamine
HexNAc	N-acetylhexosamine
HMDS	hexamethyldisilazane
HPLC	high performance liquid chromatography
HPMC	hydroxypropyl methylcellulose
IA	intraarticular
IAI	intraarticular injection
IOL	intraocular lens
IOP	intraocular pressure
IR	infrared
ISO	International Standards Organization
IU	international unit
IV	intrinsic viscosity, intravenous
LAL	Limulus Amebocyte Lysate
LC-MS	liquid chromatography-mass spectrometry
LS	light scattering
LTA	lipoteichoic acid
MALLS	multi-angle laser light scattering (photometer)
MAUDE	Manufacturer and User Device Experience database
ME	middle ear
mOsm	milliosmole
mPas	milli Pascal second
MPA	microparticle photometric agglutination
MS	mass spectrometry
MW	molecular weight
NADA	new animal drug application

NaHA	sodium hyaluronate
NASHA	Non-Animal Stabilized Hyaluronic Acid
NHS	N-hydroxysuccinimide
NMR	nucleic magnetic resonance
NSAID	non-steroid anti-inflammatory drug
OA	osteoarthritis
OME	owl monkey eye
ORD	optical rotary dispersion, oxidative reductive depolymerization
OVD	Ophthalmic Viscoelastic Device
Pas	Pascal-second
PBS	phosphate buffered solution
PEG	polyethyleneglycol
PET	positron emission tomography
PMA	premarket approval
PMN	premarket notification
PSA	postsurgical adhesion
RCT	randomized controlled trial
SAIR	severe acute inflammatory reaction
SD	standard deviation
SEC	size exclusion chromatography
SRS	(Wrinkle) Severity Rating Scale (Score)
TA	tissue augmentation
TBA	tetrabutylammonium
TEMPO	2,2,6,6-tetramethyl-1-piperidinyloxy free radical
TFUF	Tangential flow ultra filtration
TMJ	temporo-mandibular joint
TNF	tumor necrosis factor
UDP	uridine diphosphate
USP	United States Pharmacopoeia, United States Patent
UV	ultraviolet
UV-Vis	ultraviolet-visible (spectroscopy)
VAS	visual analog score
VUR	vesicoureteral reflux
WOMAC	Western Ontario and McMaster Universities (OA Index)
wt	weight

Contents

1

Overview

1.1 Discovery

Hyaluronic acid (HA), a naturally occurring linear polysaccharide, is widely known for its remarkable ability to form a highly viscoelastic solution in aqueous media. The polymer chain of HA is composed of repeating disaccharide units of glucuronic acid and N-acetylglucosamine, with a wide range of molecular weight up to several million dalton. (The name "dalton" or its symbol, Da, represents the unit of molecular mass, commonly referred to as molecular weight. Conventionally, the unit dalton is often understood.) (Fig. 1.1) HA was discovered by Meyer and Palmer in 1934, when they identified a mucoid-like substance isolated from the bovine vitreous humor as an acid polysaccharide that contains uronic acid and an amino sugar but no sulfate.[1]

The name hyaluronic is derived from hyaloid (vitreous) and uronic acid. The high–molecular weight polysaccharide was considered to be in its acid form, as it was collected by precipitation using acetic acid. Under physiological conditions, virtually all HA molecules are in their ionized form — hyaluronate. As it is not always feasible or necessary to define the counter ion (e.g., sodium, calcium) or the degree of dissociation of HA, a new nomenclature, hyaluronan, was therefore suggested as a general term irrespective of its degree of dissociation. As an acronym only for hyaluronic acid originally, HA now stands for hyaluronic acid, hyaluronate, or hyaluronan, depending on the context in which it is mentioned.[2,3] Meanwhile, a specific hyaluronate can be represented by a symbol of the metal element preceding HA. For instance, NaHA can be used to represent sodium hyaluronate.

The structural information Meyer could gather initially was quite rudimentary, if considered by the present standard. The positive results of certain key assays for acid-hydrolyzed HA identified HA as certain group or subgroup of compounds. Examples were Molisch reaction, a general colorimetric assay for carbohydrate; Elson-Morgan reaction, a colorimetric assay for amino sugar[4]; and Tollen's naphthoresorcinol reaction, a colorimetric assay for hexuronic acid. The equivalent weight of the hydrolysate was around 450,

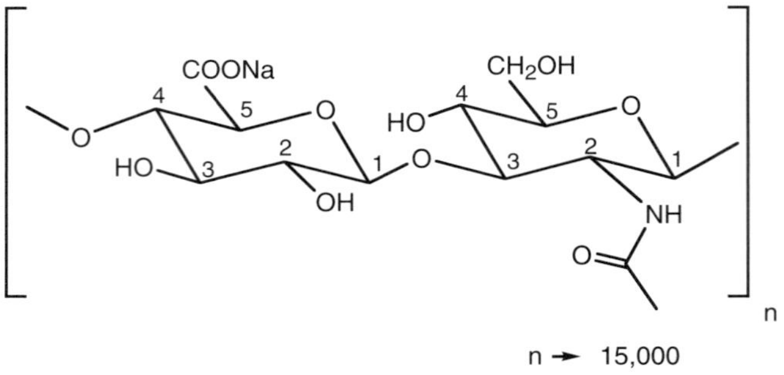

n → 15,000

FIGURE 1.1
Hyaluronic acid is a linear polysaccharide, with repeating disaccharide units of glucuronosyl-β-1,3-N-acetylglucosamine linked by β-1,4- glycosidic bonds.

as established by electro-titration using glass electrode and phenolphthalein as the indicator. The apparent dissociation constant of HA, derived from the titration data, was 4.58×10^{-5}, and HA is therefore a stronger acid than acetic acid that has a dissociation constant of 1.75×10^{-5}. The nitrogen content of HA was found in the range of 3–4%. The Kuhn and Roth method[5] indicated the existence of two acetyl groups per equivalent weight.[1] By the mid-1940s, HA had been successfully isolated from sources other than bovine eyes, such as synovial fluid, skin, umbilical cord, cock's comb, and a bacteria source — streptococci.[6]

The discovery of HA and its basic structure was achieved by applying the principles of chemistry related to HA. The process of the structural investigation of HA also improved our knowledge of its chemical behavior. The effort toward HA structural elucidation continued for many decades.

1.2   Primary Structure

To define the primary structure of HA fully, it is necessary to identify all composing monosaccharide residues, the sequence and positions of linkages of these residues, the anomeric configurations, and the positions of any substituents. Such information could be derived from the study of the building blocks of HA — its disaccharide units.

Hyaluronidases played a critical role in the early effort for structural elucidation of HA. Both testicular and bacterial hyaluronidases hydrolyze only the N-acetylglucosaminidic bonds of HA from all sources alike. Whereas testicular hyaluronidase produces tetra-, hexa-, and octasaccharides, the end product of bacterial hyaluronidase is a disaccharide (Fig. 1.2).[7] Linker and Meyer reported that the disaccharide had a strong ultraviolet absorbance at

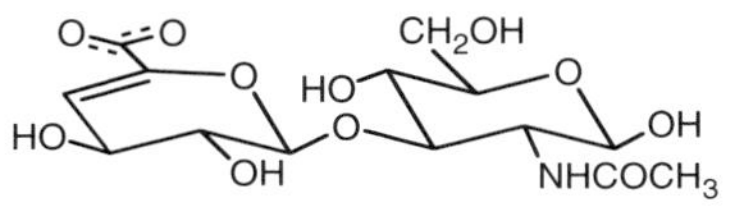

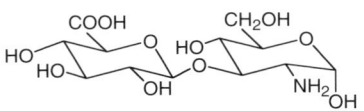

FIGURE 1.2
α-β unsaturated disaccharide identified as products from hyaluronic acid treated with bacterial hyaluronidase.

230 nm and consumed 1 M of bromine. Its hydrogenation product (homogenous by paper chromatography) no longer absorbed at 230 nm or gave positive results on uronic acid (carbazole assay) or decarboxylated on heating with acid. The Elson-Morgan reaction was positive, indicating that the aldehydic function was still intact. All these results strongly indicate that the bacterial hyaluronidase functions as a lyase that degrades HA through elimination reaction to produce an α-β unsaturated compound.[8]

The degradation products of HA by both enzymes consistently demonstrated an equal molar ratio of glucuronic acid and N-acetylglucosamine. This strongly indicated a linear structure of HA with alternating arrangement of the two sugar units.[7,9,10] The actions of enzymes are specific, and their reaction conditions are mild. HA fragments as the products of hyaluronidase treatment still maintain their basic structural features. Hyaluronidase reactions provided HA oligosaccharides, the building blocks of HA, with a size in the range of hundreds of dalton. This was far more manageable for the purpose of structural investigation than the HA polymers with molecular weight in the millions. Hyaluronidase and HA oligosaccharides, therefore, have been particularly useful in defining the detailed structure of HA.

For a complete characterization of the HA disaccharides, it was important to obtain pure chemicals with similar structure of HA disaccharides. In 1951, Meyer et al. reported in *Nature* an isolation of a crystalline disaccharide.[11] Sodium hyaluronate was hydrolyzed with purified testicular hyaluronidase and with sulfuric acid. A rectangular, long prism was obtained. Potentiometric titration indicated a substance with a pK_1 of 2.6 and pK_2 of 7.1. The zwitterionic form explained its water insolubility. The product was named hyalobiuronic acid, an N-deacetylated disaccharide of HA (Fig. 1.3). (More details about using the chemistry of hyalobiuronic acid to investigate the structure of HA will be described in Chapter 2.) Then Weissmann and Meyer

FIGURE 1.3
Hyalobiuronic acid.

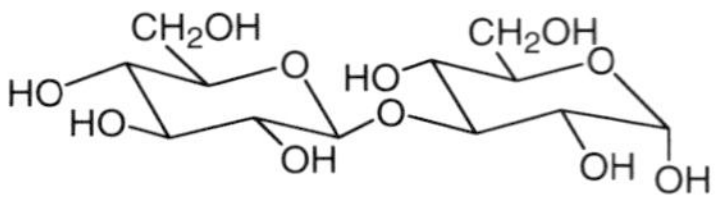

FIGURE 1.4
Laminaribiose.

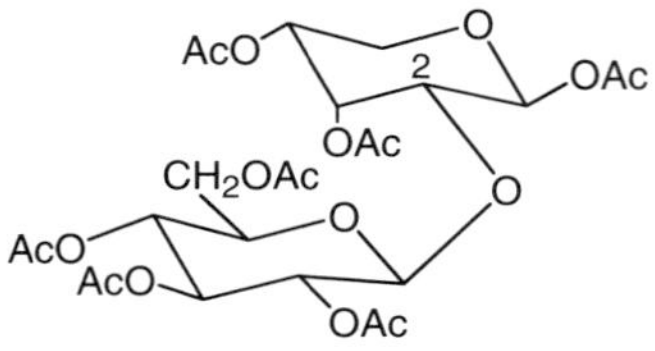

FIGURE 1.5
2-O-β-D-glucopyranosyl-D-arabinose heptaacetate.

conducted a series of chemical reactions, using hyalobiuronic acid and lam-inaribiose.[10] The latter was isolated from the hydrolysate of seaweed, the structure of which was known (Fig. 1.4).

Both disaccharides produced the same disaccharide 2-O-β-D-gluco-pyranosyl-D-arabinose heptaacetate. When the heptaacetates from the two sources were mixed, no depression of melting point (198–199°C) was observed, indicating that two heptaacetates were identical (Fig. 1.5). More detailed accounts of the chemical reactions will be provided in Chapter 2, *Chemistry*.

The information of the identity of HA disaccharide residues, the alternat-ing sequence of these residues, the positions of linkage, and the anomeric configuration allowed scientists to draw conclusions on the exact primary structure of HA: [→4)-O-(β-D-glucopyranosyluronic acid)-(1→3)-O-(2-aceta-mido-2-deoxy-β-D-glucopyranosyl)-(1→] (Fig. 1.1) Both monosaccharides have the β-D-anomeric configuration at C-1. The linkage from glucuronic acid to N-acetylglucosamine is (1→3), and the linkage from N-acetylglu-cosamine to glucuronic acid is (1→4).

1.3   Molecular Formula

The structure of HA disaccharide was often depicted by the Fischer projec-tion formulae in the early literature, such as cited above. The drawback of Fischer formulae is the difficulty to visualize in such planar drawings the spatial relations among the atoms. As an improvement over Fischer formu-lae, Haworth formulae show the orientation of each individual group relative

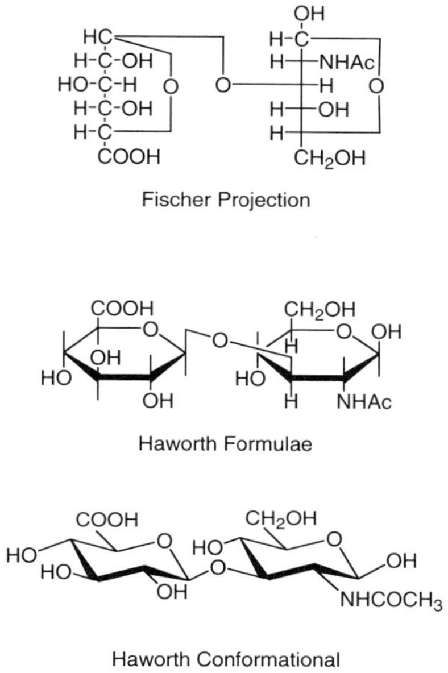

FIGURE 1.6
Molecular formula of hyaluronic acid disaccharide units.

to the imaginary ring plane. However, the visualization of the puckered ring
conformation and the configuration and position of each individual carbon
atom is best represented by Haworth "conformational" formulae (Fig. 1.6).

β-D-glucopyranose has a $^4C_1$ conformation.[12] This notation defines a ring
conformation with C-4 above the plane described by the oxygen atom and
C-2, C-3, and C-5 carbon atoms; and with C-1 below the plane. The $^4C_1$
conformation of β-D-glucopyranose is more stable than the $^1C_4$ conformation.
In a $^4C_1$ conformation, all substituents on the ring are at equatorial positions, whereas at $^1C_4$ conformation, all substituents on the ring are at axial
positions.[13]

1.4   Member of Glycosaminoglycans

HA belongs to a family of structurally similar polysaccharides known as
glycosaminoglycans (also known as mucopolysaccharides in earlier literature). Similar to HA, the names of the other polysaccharides in the group
were named after their sources — hence chondroitin sulfate (from cartilage), dermatan sulfate (from skin), keratan sulfate (from cornea), heparan
sulfate, and heparin (from liver). These polysaccharides all contain repeating

disaccharide units characterized by one amino sugar and at least one negatively charged group of carboxylate or sulfate. Chondroitin is the amino sugar C4 epimer of HA, whereas dermatan is the C5 epimer of chondroitin (Figs. 1.7 and 1.8).[12]

Heparin and heparan sulfate are collectively known as heparinoid. The repeating disaccharide units of heparinoids are more difficult to define. The complexity is the result of their structural variations. The uronic acid moiety can be D-glucuronic acid or L-iduronic acid. The D-glucosamine moiety is either N-sulfated or N-acetylated. Heparinoid may be further O-sulfated at C6 or C3 of the D-glucosamine and at C2 of the uronic acid moiety. Heparan sulfate, contrary to what its name may indicate, is actually less sulfated than heparin. Keratan sulfate, unlike the other glycosaminoglycans, contains no uronic acid residues. Sulfate is the only acid group found in keratan sulfate.

Glycosaminoglycans are the polysaccharide chains of proteoglycan. Proteoglycan and glycoprotein are macromolecules with protein backbone and pendent carbohydrate parts. There are two major differences: the carbohydrate residues of proteoglycan are often glycosaminoglycans, and the carbohydrate in proteoglycan is the dominant part, whereas protein is the dominant part in glycoprotein. Collagens, elastins, and proteoglycan are the major macromolecular components of the extracellular matrix of connective tissues.

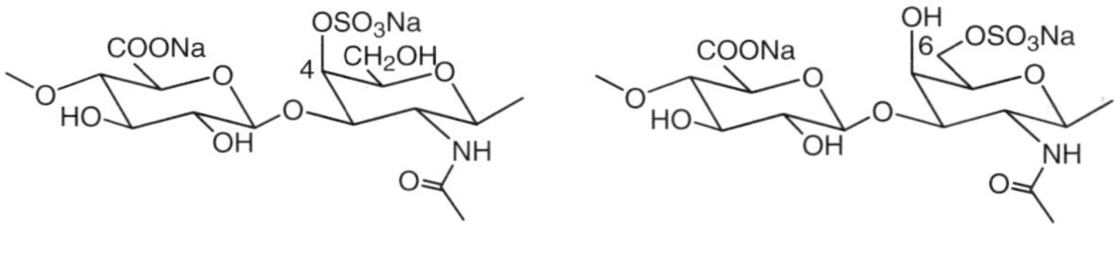

FIGURE 1.7
Chondroitin sulfates: sulfate group located on 4- and 6-positions of the acetamido-2-deoxy-D-galactosyl residues.

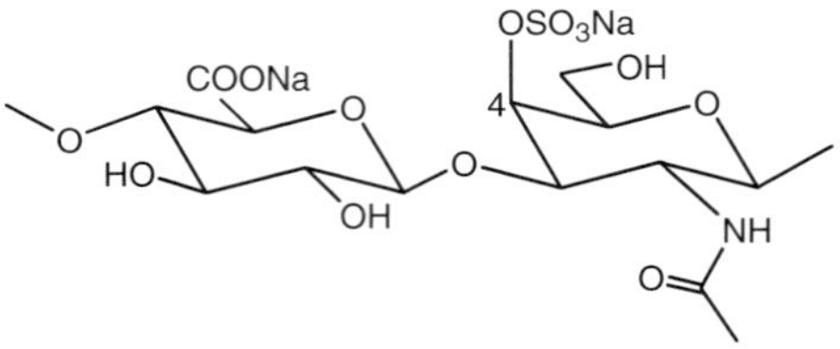

FIGURE 1.8
Dermatan sulfate, an isomer of chondroitin sulfate with L-iduronic acid residues replacing D-glucuronic acid.

1.5   HA Structures of Higher Orders

The secondary, tertiary, and quaternary structures are higher orders of structures for polymers such as polysaccharides, as compared to the primary structure mentioned above. They refer to, in large extent, the shape of an HA molecule as a whole and their intermolecular organization. Although sometimes the differences among those higher orders are not exactly clear in the literature, some basic distinctions can be made.

The secondary structure is mainly concerned with the spatial relationship of neighboring functional groups. For HA, the effect of this spatial relation is in part reflected in the rigidity of the polysaccharide chain. The degree of rigidity is largely determined by the restriction on the rotation of the glycosidic bonds by the geometry of sugar rings and, as in the case of HA, by the bulky N-acetyl group adjacent to the glycosidic bond.

For HA, the tertiary structure often refers to its molecular conformation, a geometrical pattern known as helix. Such ordered tertiary structure of HA is possible because of certain characteristics of its primary and secondary structures. The repeating disaccharide building blocks in HA chains provide regularity in sequence and spacing of functional groups. Furthermore, the helical structure is reinforced by the hydrogen bonding between the functional groups such as $C\!=\!O$, NH, and OH.

The evidence of the presence of ordered intra- and intermolecular structures of HA first came from the studies of its optical properties using polarized lights. These optical properties include optical rotations, circular dichroism (CD), and optical rotary dispersion (ORD). Whereas the absorbance and transmittance of the unpolarized lights only give information regarding the purity and concentration of HA, the optical data from the polarized lights are indicative of the existence of an ordered conformational structure of HA in solution.

Hirano and Kondo compared the specific rotation of (a) HA water solution, (b) HA dissolved in 8 M urea, and (c) HA urea reisolated by gel filtration. The authors concluded that the change in specific rotation from (a) to (b) is probably attributable to the dissociation of the HA double-stranded helix into random coil in the presence of urea.[14] ORD measures the wavelength dependence of the molecular rotation of a compound. In an ORD study, Hirano and Kondo observed at close to 600 nm (a) a negative plain curve of HA that accords with β-D configuration and (b) induced Cotton effect of the methylene blue complex with HA and heparin with opposite dip-and-rise patterns. Because heparin was assigned a right-handed helix,[14] it was suggested that the HA be assigned the left-handed helix. This was also in agreement with x-ray diffraction data of HA in its solid state.[15]

Stone reported a large negative circular dichroic band at 210 nm and an optical rotary dispersion trough at 220 nm for HA solutions,[16,17] with the observation that when pH decreased from neutral to acidic, a positive CD band appeared. Chakrabarti et al. found that at pH 2.5, the positive CD band reached maximum.[16] This phenomenon was particularly interesting, as it correlated to the simultaneous and dramatic increase of the viscosity of HA solution while the pH adjustment was made. With regard to the effect of molecular weight of HA on its solution properties, it did not come as a surprise that the molar rotation and ellipticity values of HA tetra- and hexasaccharides were much lower than those of HA polymers.

In late 1970s, data from several ^{1}H NMR (nuclear magnetic resonance) studies of HA were used to interpret certain details of the secondary/tertiary structure of the oligosaccharides of HA.[18-20] Welti et al. compared the chemical shifts of the protons on the sugar rings of HA, using D_2O as solvent, and found that the chemical shift of C2 proton of the glucuronic acid in NaOD increased by δ 0.15 compared to the HA under neutral conditions, whereas chondroitin sulfates did not have a similar change. Scott et al. used [d_6]dimethylsulfoxide as a solvent to examine the chemical shifts of the N protons. The HA protons attached to the N atom do not exchange rapidly with the deuterium in [d_6]dimethylsulfoxide; therefore, the N proton signal can be observed. The chemical shift of the N proton of the terminal N-acetyl group of HA tetrasaccharide was measured as 7.81 δ, whereas that of the internal N-acetyl group was 9.21 δ. This indicated that the internal N hydrogen was hydrogen bonded with the neighboring carboxylate, and therefore was more electron deficient than the terminal N-hydrogen. Scott et al. also found that the coupling constant of the HC-NH of the internal N-acetyl group (6.2 Hz) of HA tetrasaccharide was lower than the expected coupling constant (8.7 Hz) of the *trans* HC-NH, which indicated a change of the dihedral angle and a twist of the HA chain needed to fit the intramolecular hydrogen bonded conformation.[19] On the basis of these studies, a "super-hydrogen-bonded structure" of HA containing four different hydrogen bonds per tetrasaccharide unit was envisioned (Fig. 1.9).

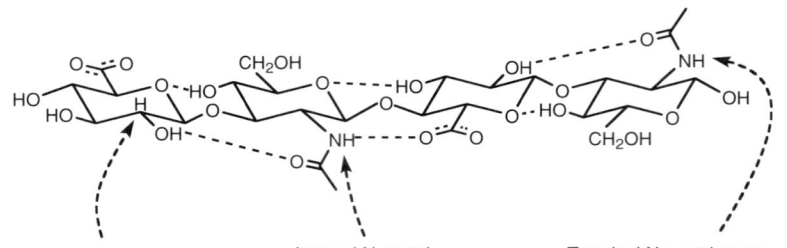

Increase of δ 0.15 in NaOD as compared with neutral condition. Chondroitin sulfates do not have the similar change

Internal N-acetyl - NH is hydrogen bonded. Chemical shift of H, 9.21δ. Solvent (d_6) DMSO.

Terminal N-acetyl group - NH is not hydrogen bonded. Chemical shift of H, 7.81δ. Solvent (d_6) DMSO.

FIGURE 1.9
Nuclear magnetic resonance study of hyaluronic acid oligosaccharide indicates a super-hydrogen-bonded structure with four different hydrogen bonds.

Using the NMR technique, Scott et al. also confirmed the formation of intramolecular HA hydrogen bonding under aqueous condition (in DMSO/water mixture).[21] The difference is, when water is present, the hydrogen bonding between NH and COO is bridged by a water molecule. This is consistent with the conclusion Scott drew from his molecular modeling study.[22]

The quaternary structure of a polysaccharide, according to Kennedy, "involves the aggregation of a number chains by noncovalent bonds."[12] For HA, its quaternary structure is related to the forming of an antiparallel double helix and a honeycomb network at extremely low concentrations of $0.5–1.0 \, \mu g/ml$. Scott concluded by computer simulation and energy calculations that this kind of helical conformation of HA is not only "sterically possible" but also "energetically likely."[22] By examining the secondary structure of HA molecules, one can find a large hydrophobic patch of eight CHs stretching across three sugar units. The hydrophobic interaction and the hydrogen bonding between the two antiparallel HA molecules are considered the driving force of such aggregation.[23] The honeycomb-like structure of HA in aqueous solution was observed under rotary shadowing electron microscope.[22]

In the higher orders (i.e., the secondary, tertiary, and quaternary structures) of HA lies the clue to many of the physical and chemical properties of HA. Chapter 3 discusses the relationship between the structural and rheological properties of HA in greater detail.

1.6   Viscosity, Molecular Weight, and Structures

HA molecules from different sources are thought to have the same primary structure. The major difference appears to be in the degree of polymerization of the HA polysaccharide chains obtained from different tissues. The bovine vitreous contains lower–molecular weight HA in the range of from 10^4 to 10^5 Da. Streptococcal cultures produce HA with molecular weight in the range of 10^5 to 10^6 Da. HA from rooster comb, human umbilical cord, and synovial fluids have higher molecular weight in the order of from 10^6 to 10^7 Da. The two key properties of HA are closely associated — its high molecular weight and large hydrodynamic volume.[24] In fact, the molecular weight data are often derived from the intrinsic viscosity data — a measure of hydrodynamic volume. The relation between the intrinsic viscosity and molecular weight is defined by the Mark-Houwink equation (Fig. 1.10).[25] A two million

$$[\eta] = K \, M^{\alpha}$$

FIGURE 1.10
Mark-Houwink equation defines the relation between intrinsic viscosity $[\eta]$ and molecular weight M. For hyaluronic acid, K is 0.036 and $\alpha = 0.78$, according to the study by T. Laurent.

molecular weight HA has an intrinsic viscosity of approximately 3000 ml/g[26] (see Table 3.1).

One should exercise caution when interpreting the data of HA molecular weight. HA isolated from animal tissues often contains various kinds of impurities, such as protein and nucleic acid. As biopolymers, protein and nucleic acid also contribute to the viscosity. The molecular weight numbers therefore usually reflect the influence of those impurities.

In the last few decades, the study of the viscoelasticity of HA has been closely associated with the investigation of its chemical structure. The electrostatic repulsion between the negative charges of the glucuronic acid appears to be an important contributing factor for the extended polymer conformation. The intra- and intermolecular hydrogen bondings and hydrophobic interactions render a certain rigidity to the HA chains. Those two structural features, as well as the large molecular weight, are recognized as the major factors contributing to the viscoelasticity of HA. Just like proteins and nucleic acids, HA gives an elegant example of the relation between structure and function in biochemistry.[27]

The usefulness of HA as medical products is dependent on their unique biophysical properties — mainly viscoelasticity and pseudoplasticity (reduction of viscosity when subject to shear force). The protective and augmenting effect of HA products on tissues is based on the ability of HA to absorb a large amount of water, fill the extracellular space, coat and lubricate the tissue surface, and absorb the mechanical stress. The fact that HA molecules from different sources have the same primary structure explains the molecular basis for its natural biocompatibility.

1.7   Source and Isolation

Human umbilical cord was the source of HA in many early researches. For HA-based products, however, more practical alternatives had to be explored and identified. HA from animal tissues such as cock's comb and HA from microbial fermentation turned out to be the two major choices for the industry.

In 1949, Boas[28] first reported the identification of HA from cock combs. "The great size of cock's comb is largely due to the presence of a thick layer of shiny mucinous connective tissue." It was known to "stain metachromatically with certain basic dyes," and therefore was believed to contain mucopolysaccharides. The isolation and purification procedure by Boas can be outlined as follows: grinding freshly harvested combs, soaking comb material in acetone to remove fat, extracting HA using sodium acetate solution, adding ethanol to precipitate HA, adding chloroform, and shaking the mixture to remove protein. The final purified dried HA was "pure white and fibrous in appearance." The yield of HA was approximately 0.1% weight of the combs processed.

The use of chloroform to remove protein was originated from Sevag,[29] who used approximately 20% chloroform to separate protein content from the carbohydrate content of the egg white. Blix and Snellman[30] adopted Sevag's method with some modifications, including increasing chloroform to 50% of the HA aqueous solution volume. Balazs[31] increased chloroform to at least equal to the HA aqueous solution during the deproteinizing process, which is a noticeable feature of the HA purification technique described in his U.S. patent of 1979.

The method of HA extraction and purification described by Swann[32] in 1968 used pronase to remove protein and used cetyl pyridinium chloride to isolate HA. The method used sufficient chloroform (not quantified) to saturate the solution. The purification was efficient. The amino acid content of the purified fraction of HA was only 0.35% w/w. A drop of HA limiting viscosity was observed that appeared to be correlated to the reduction of protein content.[33] The purified HA still maintained a molecular weight of 1.2 million. The cetyl pyridinium chloride method for purification can be referred back to the method described by Scott.[34] Long-chain quaternary ammonium salts (such as cetyl pyridinium chloride) form a complex with acidic polysaccharides (such as HA) and precipitate in aqueous media. The complex then can be dissolved and HA recovered in salt solution.

HA can also be prepared from microbial fermentation. The HA production by fermentation was first reported in 1937 by Forrest et al.[35] The method by Bracke et al.[36] produced HA from streptococcus bacteria by fermenting under anaerobic conditions. Glucose was used as the carbon source. Separation of the microorganisms from the HA is facilitated by killing the bacteria with trichloroacetic acid. Brown et al.[37,38] developed a fermentation process that does not require the killing of the microorganism under the theory that killing of microorganism may release pyrogenic substances, which would be difficult to remove. At harvest, sodium lauryl sulfate or another anionic detergent was added to the culture to release HA from the cells. Brown also chose "a special, chemically defined media" that contains no extraneous proteins. The method yielded a final HA product with low protein content.

Nimrod[39,40] developed a novel method that concerns the mutants from *Streptococcus zooepidemicus*. The microorganisms were treated with a suitable mutagen capable of producing mutants of the organism (e.g., nitrosoguanidine). The mutants lacked hemolytic activity and could produce an enhanced amount of HA. The Nimrod method was claimed to be the first fermentation method of producing high–molecular weight HA that exceeds 3.5 million.

The enzymes responsible for HA synthesis have been identified fairly recently. In 1996, several laboratories independently reported the genes encoding the hyaluronan synthases.[41] Three different HA synthases, HAS1, HAS2, and HAS3, were sequenced, and their genes, *HAS1*, *HAS2*, and *HAS3*, were designated. Spicer and McDonald[42] stated that "we are now able to convert essentially any mammalian cell, and possibly any eukaryotic cell, into a factory for hyaluronan biosynthesis simply by the expression of a single gene." The future will tell the commercial value of such technology, whether it can supply large quantity of HA at a lower cost or provide HA with unique characteristics.

1.8 Distribution and Catabolism

HA can be found extracellularly in most human tissues. Skin, lung, and intestine contain more than 50% of the HA in the body. Synovial fluid, skin, umbilical cord, and vitreous body of the eye are rich in HA content.[43,44]

The studies by Fraser et al. in the mid-1980s shed light on the catabolic pathway of HA.[45–47] The major portion of HA released from the tissues is taken up and degraded by lymphatic system. The degraded HA enters the blood and is then transported to the liver, where it is catabolized. Circulating HA is efficiently sequestered by receptor-mediated endocytosis of liver endothelial cells. Internalization of HA takes less than a minute. HA degradation in lysosomes yields glucuronic acid and N-acetylglucosamine, which leave the lysosome and are degraded to CO_2, NH_3, acetate, and lactate. These products are metabolized further by hepatocytes (parenchymal cells) to CO_2, H_2O, and urea.

Fraser et al. studied the rate of catabolism of HA in human bodies in a clinical trial using HA labeled with stable tritium at acetyl position.[48] HA was removed from the bloodstream of four normal human subjects after they received an intravenous injection of HA labeled with ^3H in the acetyl position. Identification and measuring of ^3H_2O in the urine formed the basis of an assay for HA degradation. The physiological turnover of HA was found to be remarkably rapid. The half-life of the injected HA in plasma was between 2.5 and 5.5 minutes. The daily turnover of HA in the circulation was in the range of 150–700 mg. Laurent et al. also concluded that "the daily turnover of hyaluronan is in the order of one-third of the total body content."[49,50] Such studies of the clearance of exogenous HA are particularly important from the standpoint of regulatory approval. They provide the rationale of a proposed safe dose of HA-based medical products.

The discovery that lymph nodes and liver are "responsible for the largest fraction of HA catabolism" was contrary to the prevailing thoughts at the time that HA was mostly degraded within its tissues of origin. Such a notion appeared to be based on the fact that hyaluronidase was found locally and that HA, as a macromolecule with high viscosity, would be difficult to transport to remote sites. However, no clear evidence of extracellular hyaluronidase activity was presented.[49] Hyaluronidases in most normal tissues are lysosomal enzymes that require low pH to function. For example, enzymes found in extracellular fluids like plasma[51] and synovial fluid[52] fall into such a category and therefore have no functional significance. This could be the major reason why local degradation of HA is not significant.

1.9 Influence on Cell Behavior

HA is not merely a structural component of connective tissues as was originally perceived but plays an important role in the regulation of cell behavior as well. The biophysical aspect of such a role of HA seems to be related to its ability to maintain the water balance in the extracellular matrix. The nonideal osmotic contribution of polysaccharide, as well as the flow resistance from the polymer network, give HA a major role in water homeostasis.[44,53,54]

A strong correlation between the presence of HA and cell migration and proliferation has been demonstrated. Toole found that during embryonic development, regeneration, and wound healing, the presence of the HA creates "hydrated pathways separating cellular or fibrous barriers to penetration by migrating cells."[55] Brecht et al. found that the hydrated pericellular matrices may also facilitate cell rounding during mitosis, and "inhibition of HA synthesis in fibroblasts leads to arrests of these cells in mitosis just prior to rounding," and concluded that increased HA synthesis is required for fibroblast detachment and mitosis.[56,57] Cells *in vivo* usually proliferate within an HA-rich extracellular matrix, and the presence of HA at minimum levels makes it physically possible for cells to migrate and proliferate. HA synthase activity usually increases during active division of cells and declines at confluence. In addition, Toole et al. found that high concentrations of high–molecular weight HA inhibit proliferation of fibroblasts, whereas low concentrations of low–molecular weight HA either had no effect or were somewhat stimulatory.[57,58]

The presence of HA may greatly influence the quality of wound repair. In adult wound fluid, HA peaks at 3 days and disappears at 7 days, whereas in fetus tissue repair, the early appearance and prolonged maintenance of an extracellular matrix (ECM) rich in HA is observed. Healing in fetus occurs rapidly in the absence of acute inflammation, and collagen is deposited in a highly organized, scarless manner. HA and its derivatives may have a therapeutic role in the control of the wound-healing process, because of their influence on reepithelialization, inflammation, angiogenesis, and contraction, as well as ECM synthesis, deposition, and remodeling.[59]

It is important to understand that HA regulates cell behavior in many different ways.[60] Making a general statement about the role of HA on cells without defining the specific conditions used may not be appropriate. HA formulations may vary significantly in molecular weight, purity, and concentrations, and therefore they may have different effects on various cells under specific physiological or pathological conditions.

CD44[61] and RHAMM[55,62] (receptor for hyaluronan-mediated motility) are two major classes of HA receptors. The effects of HA on cell behavior, such as migration and proliferation, are mediated by those HA receptors. The binding of HA to RHAMM triggers signal transduction, which can direct cell trafficking.[63] During cell differentiation, for example in the development of bone, lung,[64] and hair follicles, CD44 is also frequently associated with the removal of HA. One example of such morphogenetic events is the condensation of mesenchymal cells, which most likely result from partial removal of peri- and intercellular HA by CD44 mediated internalization.[55] Removal of most of the intercellular HA would eliminate the highly hydrated intercellular spaces. The residual HA then can interact with receptors on the surface of adjacent cells to form a stabilizing cross-bridging of the condensed cells.[55]

1.10 Chemical Modification

Chemical modification has been a very active area of HA research in the last 20 years. Its main purpose was to alter the chemical and physical properties of HA in a controlled way to obtain new biomaterials with certain desired features. For example, crosslinked HA hydrogels were intended to extend HA *in vivo* residence time, and HA–drug conjugates designed for sustained and targeted delivery of drugs.

Various methods of chemical modification of HA have been reported, including crosslinking using biscarbodiimide,[65] bishydrazide,[66] divinylsulfone,[67] formaldehyde,[68] bisepoxide,[69] and chemical conjugation methods using designed carbodiimides,[65] hydrazide,[70] and esterification via alkyl or benzyl halides.[71] Mono-carbodiimide alone can make HA less soluble without crosslinking.[72] The addition of multivalence metal ion (ferric chloride) to HA can also alter HA properties.[73] Refer to Chapter 2 for a detailed description of those methods.

1.11 HA-Based Medical Products

HA-based medical products approved in the U.S. are listed in Table 1.1. Invariably, HA products for various medical applications entered the market outside the U.S. before gaining approval from the Food and Drug Administration (FDA). However, the FDA review and approval documents for HA-based products are mostly available to the public and serve as a useful information source and database regarding the safety and performance of the products. Table 1.1 also lists the product trademarks and their holders, product approval dates, and categories of application. The trademark sign (™) and registered trademark sign (®) will be omitted when these products are mentioned again in the following chapters.

TABLE 1.1

Hyaluronic Acid–Based Medical Products Approved in the United States

Product Trademark	Trademark Holder[a]	Application[b]	PMA or 510K Approval Date	PMA/510K Applicant
Healon®	AMO	As OVD	P810031/1-14-83	Pharmacia
Amvisc®	B&L	As OVD	P810025/10-31-83	Med-Chem
Viscoat®	Alcon	As OVD	P840064/4-30-86	Cilco
Vitrax®	AMO	As OVD	P880031/8-10-89	Edward Weck
Provisc®	Alcon	As OVD	P890047/9-25-91	Alcon
BioLon®	Savient	As OVD	P960011/7-16-98	BTG
Hyalgan®	Fidia	IAI of OA knee	P950027/5-28-97	Fidia
Synvisc®	Genzyme	IAI of OA knee	P940015/8-8-97	Biomatrix
Supartz®	Seikagaku	IAI of OA knee	P980044/1-24-01	Seikagaku
Orthovisc®	Anika	IAI of OA knee	P030019/2-4-04	Anika
Nuflexxa™	Savient	IAI of OA knee	P010029/12-3-04	Savient
Seprafilm®	Genzyme	PSA prevention	P950034/8-12-96	Genzyme
Intergel®	Lifecore	PSA prevention	P990015/11-16-01	Lifecore
Deflux®	Q-Med	TA for VUR	P000029/9-24-01	Q-Med
Restylane®	Q-Med	TA, Cosmetic	P020023/12-12-03	Q-Med
Hylaform®	Genzyme	TA, Cosmetic	P030032/4-22-04	Genzyme
MeroGel™	Xomed	Nasal dressing	K982731/2-2-99	Xomed
MeroGel™	Xomed	Otologic dressing	K001148/7-3-00	Xomed
HylaSine®	Genzyme	Nasal dressing	K012532/10-30-01	Biomatrix

[a] Alcon Laboratories, Advanced Medical Optics (AMO), Anika Therapeutics, Bausch & Lomb (B&L), Bio-Technology General (BTG), Fidia Pharmaceutical, Lifecore Biomedical, Med-Chem Products, Q-Med AB, Savient Pharmaceuticals, Xomed Surgical.

[b] OVD: ophthalmic viscoelastic device, IAI: intraarticular injection, OA: osteoarthritis, PSA: postsurgical adhesion, TA: tissue augmentation, VUR: vesicoureteral reflux.

1.11.1 Products as Ophthalmic Surgical Aids

In the U.S., the commercial use of HA started in the ophthalmic area. It did not come as a surprise, because HA was discovered in the vitreous body of the eye and a great deal of early research was to explore the feasibility of using HA as a vitreous replacement.[74–76] The major break-through in the medical use of HA in ophthalmic area, however, occurred in cataract surgery.

In 1977, Miller et al.[77] reported that the use of HA during intraocular lens implantation in rabbits protected the corneal endothelium from damage by the plastic intraocular lens during its implantation. The eye treated with HA showed significantly less corneal edema, as well as less iris engorgement, than the eye treated without the use of HA. Then Miller and Stegmann demonstrated the protective effect of HA in human clinical studies using Healon® (HA formulation by Pharmacia) in extracapsular cataract extraction,[78] IOL implantation,[79] and other anterior segment eye surgeries.[80] In 1983, the FDA granted approval to Healon® and Amvisc® (MedChem), both from avian sources, as class III medical devices indicated for surgical aid in a variety of ophthalmic surgeries. Since then, several products containing HA from bacterial fermentation have been added to the FDA approval list, including Viscoat® (Alcon), Vitrax® (AMO), Provisc® (Alcon), and BioLon® (Bio-Tech General). Viscoat® combines HA and chondroitin sulfate in its formulation. By the end of the decade, over 80% of the cataract procedures in the U.S. were done with the aid of viscoelastics.

1.11.2   Intraarticular Injection for Knee Osteoarthritis

The clinical application of HA in the treatment of joint diseases was also much anticipated, as endogenous HA is a natural component of synovial fluid. Studies were conducted using HA to treat lameness in race horses[81] and osteoarthritis in humans knees,[82] hips,[83-85] shoulders,[86,87] and TMJ (temporo-mandibular joint),[88-91] and so forth. Five HA products for intraarticular injection to relieve the pain of osteoarthritic knee have gained FDA approval since 1996 (Table 1.1). Synvisc® (Biomatrix) is made of crosslinked HA from rooster combs. Hyalgan® (Fidia), Supartz® (Seikagaku), and Orthovisc® (Anika) are natural HA with linear structure, also derived from combs. The most recent approval by FDA (in 2004) was Nuflexxa™ (Savient), made through a fermentation process. The treatment of osteoarthritic patients with knee pain would require three to five weekly injections. The molecular weight of HA in Hyalgan® and Supartz® are below one million, and that in Orthovisc® is above one million. As a nonlinear, chemically modified HA product, the molecular weight of Synvisc® is difficult to define. One of the components of Synvisc®, hylan A, has a molecular weight of six million. The other component, hylan B, is a network, and its dimension is only limited by the size of its particles (Device Description of Synvisc, Summary of Safety and Effectiveness Data, P940015).

One theory with regard to the mode of therapeutic action was viscosupplementation, as proposed by Balazs.[92] The concept was based on the finding that in aging or arthritic joints of equine and humans, the viscosity of synovial fluid decreases. Because HA contributes most viscosity to synovial fluid, an injection of high–molecular weight HA into the joint would, as was hypothesized, restore the joint to its more normal physiological conditions.

The mechanism of action of HA in osteoarthritis therapy, however, has become a subject of debate. The opposing view argues that the molecular weight of exogenous HA applied to osteoarthritic joints does not make much difference in its therapeutic effect.[93,94] Furthermore, the intravenous administration of HA was effective in the treatment of the knee of racehorses,[95] which indicates a strong systemic effect of HA. This brings into question the significance of the local effect of a viscoelastic substance as a supplement to synovial fluid.

The controversies also exist with regard to the effectiveness of HA as a conservative therapy for osteoarthritis. The human clinical studies have shown conflicting results, probably because of the variability in the protocols and the material used in those studies. Lo et al. conducted a metaanalysis[96] on this subject, including 18 randomized, blind controlled studies between 1986 and 2002. The study concluded that intraarticular injection of HA has "at best, a small effect in the treatment of knee OA when compared with placebo. . . . The presence of publication bias suggests this effect is overestimated."[96] At the same time, the analysis could not establish whether higher–molecular weight HA has a greater efficacy than lower–molecular weight HA. The basic conclusion was reaffirmed by an expanded metaanalysis by the same group of researchers.[97] The controversy is also reflected in the different opinions of the two major organizations in the osteoarthritis field. Whereas the American College of Rheumatology made a more definitive recommendation on the efficacy of HA therapy; the European League of Associations of Rheumatologists is circumspect and only rated intraarticular injection of HA as "probably effective in knee OA."[98] The European League of Associations of Rheumatologists is known for taking evidence-based approach in their recommendations.

Despite the controversy, HA therapy for knee osteoarthritis appears to be holding its commercial ground. Synvisc, the major product for OA indication in the U.S., reached a total worldwide sale of over nine million doses by 2003 (Hylaform Panel Meeting Minutes: United States of America, Food and Drug Administration, Center for Devices and Radiological Health, Medical Devices Advisory Committee, General and Plastic Surgery Devices Panel, 64th Meeting, November 21, 2003, Gaithersburg, Maryland). It is conceivable that there may exist subgroups of the patient population that tend to respond favorably to viscoelastic HA injection. However such subgroups, if existing, have not been clearly identified. The ongoing debate remains interesting, as it stimulates our thinking and improves the basic science of HA. Good science is likely to lead to good products, and the clinical evidence will be the final arbiter.

1.11.3 Prevention of Postsurgical Adhesion

The third major medical application of HA has been in the area of prevention of postsurgical adhesion. As known from the lesson of the wound healing of fetus after surgery, scarless healing closely correlates to the presence of large amount of HA. Most HA products in this area of application are chemically modified to increase the residence time. The presence of HA at the site of surgery in the first couple of weeks is critical in controlling the fibrosis. Seprafilm® (Genzyme) is a polymer film made of chemically modified HA and carboxymethylcellulose.[99] It received approval by the FDA in 1996 and was "indicated for use in patients undergoing abdominal or pelvic laparotomy as an adjunct intended to reduce the incidence, extent, and severity of postoperative adhesions." The mechanism of action is believed to be a physical barrier separating the apposing surfaces of the damaged peritoneal during the early stage of wound repair.[100] The effect of such a solid barrier product, however, may be limited only to the sites where HA is applied. Significant postsurgical adhesions may be *de novo*, away from the surgical sites, and cannot be exactly predicted. Hence, the rationale for the "lavage" approach; that is, the use of large volumes of diluted solution to cover all tissue surface areas. Intergel® of Lifecore Biomedical,[101] a 0.5% ferric hyaluronate dilute solution, is a "lavage" type of product that gained FDA approval in 2001. Another lavage product, Sepracoat™ of Genzyme, made of 0.4% of natural HA, did not pass the scrutiny of FDA panel because of the product's lack of efficacy. The premarket approval application of Sepracoat™ by Genzyme Corporation, P960003, was voted by the General and Plastic Surgery Devices Panel as not approvable on May 5, 1997.

In addition, HA-containing products have extended their use to the nasal and otologic area. MeroGel™, made by Fidia and generically known as HYAFF — a family of esterified HA derivatives — obtained marketing clearance in the U.S. as a space-occupying dressing in nasal/sinus cavities and in the middle ear and external ear canal. HylaSine®, a crosslinked HA gel (hylan B), obtained marketing clearance in the U.S. as a space-occupying gel stent to separate mucosal surfaces in the nasal cavity. HylaSine® products serve to occupy the nasal or otologic spaces, separate mucosal surfaces, prevent adhesion, and control bleeding following surgeries. Notably, these products, also regulated as medical devices, required only 510 K filing, a regulatory approval process less demanding than that of a premarket approval application. At the other end of the spectrum of regulatory requirement, CTX-100 of Exhale Therapeutics (Cotherix), an aerosolized HA product to protect lung elastin fibers and slow the progression of emphysema, is being regulated by the FDA as a drug. The product approval of a drug is in general a long process that requires comprehensive data from multiple phases of studies.

1.11.4   Cosmetic Tissue Augmentation

The use of HA as cosmetic implant to augment tissue and reduce wrinkles is a very active area of HA product development. Unlike the case of HA for knee injection, there is little doubt about the relevance of the *in vivo* half-life of a cosmetic tissue implant to its clinical success. For that reason, Restylane® (Q-Med) and Hylaform® (Genzyme), two HA injection products for wrinkle treatment, are both based on the chemical crosslinking of HA.

Both products were considered as safe and effective as the collagen-based product already on the market — Zyplast® (Inamed Aesthetics) — although the proper balance between the change of the physical properties of modified HA and maintaining the minimum tissue reaction remains a challenge.[102–107]

1.11.5   Bone and Dental Repair

High–molecular weight HA can stimulate osteoinduction.[108] HA can be used as a carrier for demineralized, freeze-dried allograft bone particles. The mixture forms a thick paste and is used to treat bony defects with significantly better results than the saline control.[109]

The dental application of HA is being investigated as well. In a clinical study, HA derivative HYAFF first served as a scaffolding material for autologous fibroblast cells, and the graft was then used for gingival augmentation surgery.[110]

The information of most of the products mentioned in the overview, especially their compositions, safety, and performance, will be presented and discussed in greater details in Chapters 5 and 6. The general analytical methods and standards applied to HA-based medical products will be discussed in Chapter 4.

1.12   Ever-Growing Interest

The growing interest and activity in HA-related research and development are clearly indicated by the ever-increasing numbers of publications per year related to the subject. According to the Pub-Med database, the annual number of such publications was only 36 in 1965 but exceeded 100 in 1968, 200 in 1980, 300 in 1989, 400 in 1994, 500 in 1999, 600 in 2001, and 700 in 2004 (search term: hyaluronan OR hyaluronic OR hyaluronate; Fig. 1.11).

A similar trend has occurred in the area of HA-related intellectual properties. From 1979 to 1988, the U.S. issued approximately five HA-related patents per year. The annual average number surged past 35 in 1988 and maintained at a relatively close range from 1988 to 1995, when the number exceeded 50. Another sharp rise in 1996 pushed the number to around 100. The pace of issuing HA patents in the U.S. has been kept at or above 100 per year since then, with no sign of abating.

As the title of this book would suggest, the emphasis is on the practical aspects related to the development of HA-based medical products — their chemistry, composition, formulation, testing, safety, effectiveness, quality control, and regulatory approval. The subject of structure property relationship

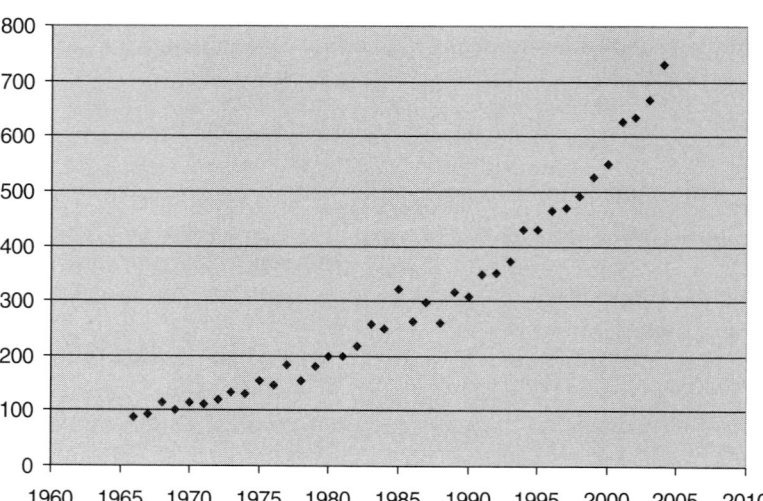

FIGURE 1.11
Steady increase of publications of hyaluronic acid–related research, development, and clinical applications in the last four decades — a clear trend of continuing interest in hyaluronic acid from academia and industry as well.

of HA is basic, important, and always intriguing. It will serve as a common thread throughout this book.

References

1. Meyer, K. and Palmer, J., The polysaccharide of the vitreous humor, *J. Biol. Chem.*, 107, 629–634 (1934).
2. Laurent, T., The biology of hyaluronan. Introduction. *Ciba Found Symp.*, 143, 1–20 (1989).
3. Balazs, E. A., Laurent, T. C., and Jeanloz, R. W., Nomenclature of hyaluronic acid, *Biochem. J.* 235, 903 (1986).
4. Elson, L. A. and Morgan, W. T. J., A colorimetric method for the determination of glucosamine and chondrosamine, *Biochem. J.*, 27, 1824–1828 (1933).
5. Kuhn, R. and Roth, H., Mikro-Bestimmung von Acetyl-, Benzoyl- und C-Methylgruppen, *Ber. Chem. Ges.*, 66, 1274–1278 (1933).
6. Meyer, K., The biological significance of hyaluronic acid and hyaluronidase, *Physiol. Revs.*, 27, 335–359 (1947).
7. Rapport, M. M., Linker, A., and Meyer, K. The hydrolysis of hyaluronic acid by pneumococcal hyaluronidase, *J. Biol. Chem.*, 192, 283–291 (1951).

8. Linker, A. and Meyer, K., Production of unsaturated uronides by bacterial hyaluronidases, *Nature*, 174, 1192–1194 (1954).
9. Rapport, M. M., Meyer, K., and Linker, A., Analysis of the products formed on hydrolysis of hyaluronic acid by testicular hyaluronidase, *J. Am. Chem. Soc.*, 73, 2416–2420 (1951).
10. Weissman, B. and Meyer, K., The structure of hyalobiuronic acid and of hyaluronic acid from umbilical cord, *J. Am. Chem. Soc.*, 76, 1753–1757 (1954).
11. Rapport, M. M., Weissman, B., Linker, A., and Meyer, K., Isolation of a crystalline disaccharide, hyalobiuronic acid, from hyaluronic acid, *Nature*, 168, 996–997 (1951).
12. Kennedy, J. F. and White, C. A., *Bioactive Carbohydrates in Chemistry, Biochemistry, and Biology*, Ellis Horwood, New York, 1983.
13. IUPAC-IUB Joint Commission on Biochemical Nomenclature (JCBN), Conformational nomenclature for five and six membered ring forms of monosaccharides and their derivatives, *Eur. J. Biochem.*, 111, 295–298 (1980).
14. Hirano, S. and Kondo, S., Molecular conformational transition of hyaluronic acid in solution, *J. Biochem.*, 74, 861–862 (1973).
15. Dea, I. C. M., Moorhouse, R., Rees, D.A., Arnott, S., Guss, J.M., and Balazs, E.A., Hyaluronic acid: a novel, double helical molecule, *Science*, 179, 560–562 (1973).
16. Chakrabarti, B. and Balazs, E. A., Optical properties of hyaluronic acid, ultraviolet circular dichroism and optical rotatory dispersion. *J. Mol. Biol.*, 78, 135–141 (1973).
17. Stone, A. L., Optical rotary dispersion of mucopolysaccharide and mucopolysaccharide-dye complex II. Ultraviolet cotton effects in the amide transition bands, *Biopolymers*, 7, 173–188 (1969).
18. Welti, D., Rees, D. A., and Welsh, E. J., Solution conformation of glycosaminoglycans: assignment of the 300-MHz 1H-magnetic resonance spectra of chondroitin 4-sulphate, chondroitin 6-sulphate and hyaluronate, and investigation of an alkali-induced conformation change, *Eur. J. Biochem.*, 94, 505–514 (1979).
19. Scott, J. E., Heatley, F., and Hull, W. E., Secondary structure of hyaluronate in solution. A 1H-N.M.R. investigation at 300 and 500 MHz in [2H6]dimethyl sulphoxide solution, *Biochem. J.*, 220, 197–205 (1984).
20. Bociek, S. M., Darke, A. H., Welti, D., and Rees, D. A., The 13C-NMR spectra of hyaluronate and chondroitin sulphates. Further evidence on an alkali-induced conformation change, *Eur. J. Biochem.*, 109, 447–456 (1980).
21. Heatley, F. and Scott, J. E., A water molecule participates in the secondary structure of hyaluronan, *Biochem. J.*, 254, 489–493 (1988).
22. Scott, J. E., Cummings, C., Brass, A., and Chen, Y., Secondary and tertiary structures of hyaluronan in aqueous solution, investigated by rotary shadowing-electron microscopy and computer simulation. Hyaluronan is a very efficient network-forming polymer, *Biochem. J.*, 274, 699–705 (1991).
23. Scott, J. E., Chemical morphology of hyaluronan, in *The Chemistry, Biology, and Medical Applications of Hyaluronan and Its Derivatives*, Laurent, T. C., Ed., Portland Press, London, 1998, pp. 7–15.
24. Morris, E. R., Rees, D. A., and Welsh, E. J., Conformation and dynamic interactions in hyaluronate solutions, *J. Mol. Biol.*, 138, 383–400 (1980).
25. Laurent, T. C., Ryan, M., and Pletruszkiewicz, A., Fractionation of hyaluronic acid, the polydispersity of hyaluronic acid from the bovine vitreous body, *Biochim. Biophys. Acta* 42, 476–485 (1960).
26. Leith, M. M., Loftus, S. A., Kuo, J. W., DeVore, D. P., and Keates, E. U., Comparison of the properties of AMVISC and Healon, *J. Cataract Refract. Surg.*, 13, 534–536 (1987).

27. Caplan, A. I., Cartilage, *Sci. Am.,* 251, 84–87, 90–94 (1984).
28. Boas, N., Isolation of hyaluronic acid from the cock's comb, *J. Biol. Chem.,* 181, 573–575 (1949).
29. Sevag, M. G., Eine neue physikalische Enteiweissungsmethode zur Darstellung biologisch wirksamer Substanzen. Isolierung von Kohlenhydraten aus Huehnereweiss und Pneumococcen, *Biochem. Zeitschrift,* 273, 419–429 (1934).
30. Blix, G. and Snellman, O., On chondroitin sulphuric acid and hyaluronic acid, *Arkiv. Kemi.,* 19A, 1–19 (1945).
31. Balazs, E. A., Ultrapure hyaluronic acid and the use thereof, U.S. Patent 4,141,973, (1979).
32. Swann, D. A., Studies on hyaluronic acid. I. The preparation and properties of rooster comb hyaluronic acid. *Biochim. Biophys. Acta,* 156, 17–30 (1968).
33. Swann, D. A., Studies on hyaluronic acid. II. The protein component(s) of rooster comb hyaluronic acid, *Biochim. Biophys. Acta,* 160, 96–105 (1968).
34. Scott, J. E., Aliphatic ammonium salts in the assay of acidic polysaccharides from tissues, *Meth. Biochem. Anal.,* 8, 145–197 (1960).
35. Forrest, E. K., Heidelberger, M., and Dawson, M. H., A serologically inactive polysaccharide elaborated by mucoid strains of group A hemolytic Streptococcus, *J. Biol. Chem.,* 118, 61–69 (1937).
36. Bracke, J. W. and Thacker, K., Hyaluronic acid from bacterial culture, U.S. Patent 4,517,295 (1985).
37. Brown, K. K., Ruiz, L. L. C., and Van De Rijn, I., Ultrapure hyaluronic acid and method of making it, U.S. Patent 4,782,046 (1988).
38. Brown, K. K., Ruiz, L. L. C., Van De Rijn, I., Greene, N. D., Trump, S. L., Wilson, C. D., and Bryant, S. A., Method for the microbiological production of non-antigenic hyaluronic acid, U.S. Patent 5,316,926 (1994).
39. Nimrod, A., Greenman, B., Kanner, D., Landsberg, M., and Beck, Y., Method of producing high molecular weight sodium hyaluronate by fermentation of streptococcus, U.S. Patent 4,780,414 (1988).
40. Nimrod, A., Greenman, B., Kanner, D., and Landsberg, M., High molecular weight sodium hyaluronate, U.S. Patent 4,784,990 (1988).
41. Weigel, P. H., Hascall, V. C., and Tammi, M., Hyaluronan synthases, *J. Biol. Chem.,* 272, 13997–14000 (1997).
42. Spicer, A. P. and McDonald, J. A., Eukaryotic hyaluronan synthases, www.glycoforum.gr.jp/science/hyaluronan, 1998.
43. Reed, R. K. and Laurent, U. B., Turnover of hyaluronan in the microcirculation, *Am. Rev. Respir. Dis.,* 146, S37–S39 (1992).
44. Laurent, T. C., in *Chemistry and Molecular Biology of the Intercellular Matrix,* Balazs, E. A., Ed., Academic Press, London, 1970, pp. 703–732.
45. Laurent, T. C., Dahl, I. M., Dahl, L. B., Engstrom-Laurent, A., Eriksson, S., Fraser, J. R., Granath, K. A., Laurent, C., Laurent, U. B., Lilja, K., The catabolic fate of hyaluronic acid, *Connect. Tissue Res.,* 15, 33–41 (1986).
46. Fraser, J. R., Kimpton, W. G., Laurent, T. C., Cahill, R. N., and Vakakis, N., Uptake and degradation of hyaluronan in lymphatic tissue, *Biochem. J.,* 256, 153–158 (1988).
47. Smedsrod, B., Cellular events in the uptake and degradation of hyaluronan, *Adv. Drug Delivery Rev.,* 7, 265–278 (1991).
48. Fraser, J. R., Laurent, T. C., Engstrom-Laurent, A., and Laurent, U. G., Elimination of hyaluronic acid from the blood stream in the human, *Clin. Exp. Pharmacol. Physiol.,* 11, 17–25 (1984).

49. Fraser, J. R. E., Brown, T. J., and Laurent, T. C., Catabolism of hyaluronan, in *The Chemistry, Biology, and Medical Applications of Hyaluronan and Its Derivatives*, Laurent, T. C., Ed., Portland Press, London, 1998, pp. 85–92.
50. Laurent, T. C. and Fraser, J. R. E., in *Degradation of Bioactive Substances: Physiology and Pathophysiology*, Henriksen, J. H., Ed., CRC Press, Boca Raton, FL, 1991, pp. 249–265.
51. De Salegui, M. and Pigman, W., The existence of an acid-active hyaluronidase in serum, *Arch. Biochem. Biophys.*, 120, 60–67 (1967).
52. Stephens, R. W., Ghosh, P., and Taylor, T. K., The characterisation and function of the polysaccharidases of human synovial fluid in rheumatoid and osteoarthritis, *Biochim. Biophys. Acta*, 399, 101–112 (1975).
53. Laurent, T. C. and Fraser, J. R., Hyaluronan, *FASEB J.*, 6, 2397–2404 (1992).
54. Comper, W. D. and Laurent, T. C., Physiological function of connective tissue polysaccharides, *Physiol. Rev.*, 58, 255–315 (1978).
55. Toole, B., Hyaluronan-cell interactions in morphogenesis, in *The Chemistry, Biology, and Medical Applications of Hyaluronan and Its Derivatives*, Laurent, T. C., Ed., Portland Press, London, 1998, pp. 155–160.
56. Brecht, M., Mayer, U., Schlosser, E., and Prehm, P., Increased hyaluronate synthesis is required for fibroblast detachment and mitosis, *Biochem. J.*, 239, 445–450 (1986).
57. Toole, B., in *Cell Biology of Extracellular Matrix*, Hay, E. D., Ed., Plenum, New York, 1991, pp. 305–341.
58. Goldberg, R. L. and Toole, B. P., Hyaluronate inhibition of cell proliferation, *Arthritis Rheum.*, 30, 769–778 (1987).
59. Adzick, N. S. and Longaker, M. T., Scarless fetal healing. Therapeutic implications, *Ann. Surg.*, 215, 3–7 (1992).
60. Tammi, M. I., Day, A. J., and Turley, E. A., Hyaluronan and homeostasis: a balancing act, *J. Biol. Chem.*, 277, 4581–4584 (2002).
61. Aruffo, A., Stamenkovic, I., Melnick, M., Underhill, C., and Seed, B., CD44 is the principal cell surface receptor for hyaluronate, *Cell*, 61, 1303–1313 (1990).
62. Hardwick, C., Hoare, K., Owens, R., Hohn, H. P., Hook, M., Moore, D., Cripps, V., Austen, L., Nance, D. M., and Turley, E. A., Molecular cloning of a novel hyaluronan receptor that mediates tumor cell motility, *J. Cell Biol.*, 117, 1343–1350 (1992).
63. Entwistle, J., Hall, C. L., and Turley, E. A., HA receptors: regulators of signalling to the cytoskeleton, *J. Cell Biochem.*, 61, 569–77 (1996).
64. Underhill, C. B., Nguyen, H. A., Shizari, M., and Culty, M., CD44 positive macrophages take up hyaluronan during lung development, *Dev. Biol.*, 155, 324–336 (1993).
65. Kuo, J. W., Swann, D. A., and Prestwich, G. D., Chemical modification of hyaluronic acid by carbodiimides, *Bioconjug. Chem.*, 2, 232–241 (1991).
66. Luo, Y., Kirker, K. R., and Prestwich, G. D., Crosslinked hyaluronic acid hydrogel films: new biomaterials for drug delivery, *J. Control Release*, 69, 169–184 (2000).
67. Balazs, E. A. and Leshchiner, A., Cross-linked gels of hyaluronic acid and products containing such gels, U.S. Patent 4,582,865 (1986).
68. Balazs, E. A., Leshchiner, A., Leshchiner, A., and Band, P., Chemically modified hyaluronic acid preparation and method of recovery thereof from animal tissues, U.S. Patent 4,713,448 (1987).
69. Malson, T. and Lindqvist, B. L., Gel of crosslinked hyaluronic acid for use as a vitreous humor substitute, U.S. Patent 4,716,154 (1987).

70. Pouyani, T. and Prestwich, G. D., Functionalized derivatives of hyaluronic acid oligosaccharides: drug carriers and novel biomaterials, *Bioconjug. Chem.*, 5, 339–347 (1994).
71. Della Valle, F. and Romeo, A., Esters of hyaluronic acid, U.S. Patent 4,851,521 (1989).
72. Hamilton, R., Fox, E. M., Acharya, R. A., and Walts, A. E., Water insoluble derivatives of hyaluronic acid, U.S. Patent 4,937,270 (1990).
73. Huang, W. J., Johns, D. B., and Kronenthal, R. L., Ionically crosslinked carboxyl-containing polysaccharides for adhesion prevention, U.S. Patent 5,532,221, lifecore (1996).
74. Pruett, R. C., Schepens, C. L., and Swann, D. A., Hyaluronic acid vitreous substitute. A six-year clinical evaluation, *Arch. Ophthalmol.*, 97, 2325–2330 (1979).
75. Balazs, E. A. and Sweeney, D. B., in *New and Controversial Aspects of Retinal Detachment*, McPherson, A., Ed., Harper Row, New York, 1968, pp. 371–376.
76. Goa, K. L. and Benfield, P., Hyaluronic acid. A review of its pharmacology and use as a surgical aid in ophthalmology, and its therapeutic potential in joint disease and wound healing, *Drugs*, 47, 536–566 (1994).
77. Miller, D., O'Connor, P., and Williams, J., Use of Na-hyaluronate during intraocular lens implantation in rabbits, *Ophthal. Surg.*, 8, 58–61 (1977).
78. Stegmann, R. and Miller, D., Extracapsular cataract extraction with hyaluronate sodium, *Ann. Ophthalmol.* 14, 813–815 (1982).
79. Miller, D. and Stegmann, R., Use of sodium hyaluronate in human IOL implantation, *Ann. Ophthalmol.*, 13, 811–815 (1981).
80. Miller, D. and Stegmann, R., Use of Na-hyaluronate in anterior segment eye surgery, *J. Am. Intraocul. Implant Soc.*, 6, 13–15 (1980).
81. Butler, J., Rydell, N. W., and Balazs, E. A., Hyaluronic acid in synovial fluid. VI. Effect of intra-articular injection of hyaluronic acid on the clinical symptoms of arthritis in track horses, *Acta Vet. Scand.*, 11, 139–155 (1970).
82. Peyron, J. G. and Balazs, E. A., Preliminary clinical assessment of Na-hyaluronate injection into human arthritic joints, *Pathol. Biol. (Paris)*, 22, 731–736 (1974).
83. Migliore, A., Martin, L. S., Alimonti, A., Valente, C., and Tormenta, S., Efficacy and safety of viscosupplementation by ultrasound-guided intra-articular injection in osteoarthritis of the hip, *Osteoarthritis Cartilage*, 11, 305–306 (2003).
84. Brocq, O., Tran, G., Breuil, V., Grisot, C., Flory, P., and Euller-Ziegler, L., Hip osteoarthritis: short-term efficacy and safety of viscosupplementation by hylan G-F 20. An open-label study in 22 patients, *Joint Bone Spine*, 69, 388–391 (2002).
85. Lust, G., Beilman, W. T., Dueland, D. J., and Farrell, P. W., Intra-articular volume and hip joint instability in dogs with hip dysplasia, *J. Bone Joint Surg. Am.*, 62, 576–582 (1980).
86. Itokazu, M. and Matsunaga, T., Clinical evaluation of high-molecular-weight sodium hyaluronate for the treatment of patients with periarthritis of the shoulder, *Clin. Ther.*, 17, 946–955 (1995).
87. Leardini, G., Perbellini, A., Franceschini, M., and Mattara, L., Intra-articular injections of hyaluronic acid in the treatment of painful shoulder, *Clin. Ther.*, 10, 521–556 (1988).
88. Bertolami, C. N., Gay, T., Clark, G. T., Rendell, J., Shetty, V., Liu, C., and Swann, D. A., Use of sodium hyaluronate in treating temporomandibular joint disorders: a randomized, double-blind, placebo-controlled clinical trial, *J. Oral Maxillofac. Surg.*, 51, 232–242 (1993).

89. McCain, J. P., Balazs, E. A., and de la Rua, H., Preliminary studies on the use of a viscoelastic solution in arthroscopic surgery of the temporomandibular joint, *J. Oral Maxillofac. Surg.*, 47, 1161–1168 (1989).

90. Kopp, S., Carlsson, G. E., Haraldson, T., and Wenneberg, B., Long-term effect of intra-articular injections of sodium hyaluronate and corticosteroid on temporomandibular joint arthritis, *J. Oral Maxillofac. Surg.*, 45, 929–935 (1987).

91. Kopp, S., Wenneberg, B., Haraldson, T., and Carlsson, G. E., The short-term effect of intra-articular injections of sodium hyaluronate and corticosteroid on temporomandibular joint pain and dysfunction, *J. Oral Maxillofac. Surg.*, 43, 429–435 (1985).

92. Balazs, E. A. and Denlinger, J. L., Viscosupplementation: a new concept in the treatment of osteoarthritis, *J. Rheumatol. Suppl.*, 39, 3–9 (1993).

93. Brandt, K. D., Smith, G. N. Jr., and Simon, L. S., Intraarticular injection of hyaluronan as treatment for knee osteoarthritis: what is the evidence?, *Arthritis Rheum.*, 43, 1192–1203 (2000).

94. Aviad, A. D. and Houpt, J. B., The molecular weight of therapeutic hyaluronan (sodium hyaluronate): how significant is it?, *J. Rheumatol.*, 21, 297–301 (1994).

95. Kawcak, C. E., Frisbie, D. D., Trotter, G. W., McIlwraith, C. W., Gillette, S. M., Powers, B. E., and Walton, R. M., Effects of intravenous administration of sodium hyaluronate on carpal joints in exercising horses after arthroscopic surgery and osteochondral fragmentation, *Am. J. Vet. Res.*, 58, 1132–1140 (1997).

96. Lo, G. H., La Valley, M. P., McAlindon, T. E., and Felson, D. T., Is intra-articular (IA) hyaluronic acid (HA) efficacious in treating knee osteoarthritis (OA)? A meta-analysis, in *66th Annual Scientific Meeting of the ACR (American College of Rheumatology)*, 2002.

97. Lo, G., LaValley, M., McAlindon, T., and Felson, D., Intra-articular hyaluronic acid in treatment of knee osteoarthritis: a meta-analysis, *JAMA*, 290, 3115–3121 (2003).

98. Mazières, B., Bannwarth, B., Dougados, M., and Lequesne, M., EULAR recommendations for the management of knee osteoarthritis. Report of task force of the Standing Committee for International Clinical Studies Including Therapeutic Trials, *Joint Bone Spine*, 68, 231–240 (2001).

99. Diamond, M. P., Reduction of adhesions after uterine myomectomy by Seprafilm membrane (HAL-F): a blinded, prospective, randomized, multicenter clinical study. Seprafilm Adhesion Study Group, *Fertil. Steril.*, 66, 904–910 (1996).

100. Shimanuki, T., Nishimura, K., Montz, F. J., Nakamura, R. M., and diZerega, G. S., Localized prevention of postsurgical adhesion formation and reformation with oxidized regenerated cellulose, *J. Biomed. Mater. Res.*, 21, 173–185 (1987).

101. Johns, D. B., Keyport, G. M., Hoehler, F., and diZerega, G. S., Reduction of postsurgical adhesions with Intergel adhesion prevention solution: a multicenter study of safety and efficacy after conservative gynecologic surgery, *Fertil. Steril.*, 76, 595–604 (2001).

102. Sclafani, A. P. and Romo, T., III, Injectable fillers for facial soft tissue enhancement, *Facial Plast. Surg.*, 16, 29–34 (2000).

103. Bergeret-Galley, C., Latouche, X., and Illouz, Y. G., The value of a new filler material in corrective and cosmetic surgery: DermaLive and DermaDeep, *Aesthetic Plast. Surg.*, 25, 249–255 (2001).

104. Lupton, J. R. and Alster, T. S., Cutaneous hypersensitivity reaction to injectable hyaluronic acid gel, *Dermatol. Surg.*, 26, 135–137 (2000).

105. Shafir, R., Amir, A., and Gur, E., Long-term complications of facial injections with Restylane (injectable hyaluronic acid), *Plast. Reconstr. Surg.*, 106, 1215–1216 (2000).

106. Manna, F., Dentini, M., Desideri, P., De Pita, O., Mortilla, E., and Maras, B., Comparative chemical evaluation of two commercially available derivatives of hyaluronic acid (Hylaform from rooster combs and Restylane from streptococcus) used for soft tissue augmentation, *J. Eur. Acad. Dermatol. Venereol,* 13, 183–192 (1999).
107. Duranti, F., Salti, G., Bovani, B., Calandra, M., and Rosati, M. L., Injectable hyaluronic acid gel for soft tissue augmentation. A clinical and histological study, *Dermatol. Surg.,* 24, 1317–1325 (1998).
108. Sasaki, T. and Watanabe, C., Stimulation of osteoinduction in bone wound healing by high-molecular hyaluronic acid, *Bone.* 16, 9–15 (1995).
109. Gertzman, A. A. and Sunwoo, M. H., A pilot study evaluating sodium hyaluronate as a carrier for freeze-dried demineralized bone powder, *Cell Tissue Banking,* 2, 87–94 (2001).
110. Prato, G. P., Rotundo, R., Magnani, C., Soranzo, C., Muzzi, L., and Cairo, F., An autologous cell hyaluronic acid graft technique for gingival augmentation: a case series, *J. Periodontol.,* 74, 262–267 (2003).

2

Chemistry

2.1 Chemistry in Structural Investigation

The full characterization of hyaluronic acid (HA) in the early years of its discovery was a rather difficult and tedious task, because of the lack of sophisticated analytical tools such as nuclear magnetic resonance (NMR), infrared (IR) and efficient separation techniques such as high-performance liquid chromatography (HPLC). Defining the structure of HA depended, to a large extent, on the data and information from the chemical behavior of HA and the chemical synthesis of HA from structurally similar compounds.

Two of the major challenges to chemists in their studies of HA are the huge size of the HA molecule and its low solubility in organic solvents. The size of HA may be reduced in a controlled way by enzymatic degradation or hydrolysis. Derivatization of HA with small organic chemicals such as alcohol and anhydride can turn HA into its lipophilic derivatives that are soluble in organic solvents (e.g., benzene, toluene, etc.). This facilitates the handling of the reagents and the purification of the reaction products. However, in most cases, an expedient and practical way to conduct chemical reactions with HA is to identify suitable chemical reagents and reactions that are compatible with aqueous media.

2.1.1 Structure Verification via Chemical Reactions

Chapter 1 mentioned an early example of using series of chemical reactions to verify the structure of HA by Weissmann and Meyer in 1950s.[1] The chemical reactions started from two known molecules, hyalobiuronic acid and laminaribiose, with structures similar to HA disaccharide. Through a few steps of reactions using well-defined chemical methods (mostly acylation of the hydroxyl groups and the alkylation of the carboxyl groups), both starting carbohydrates afforded the identical heptaacetate disaccharide (Fig. 2.1).

Another example was the work by Jeanloz and Flower[2,3] in the early 1960s. They studied the degradation of HA by methanolysis, followed by total acetylation. The resultant heptaacetyl-methylester compound was well crystallized,

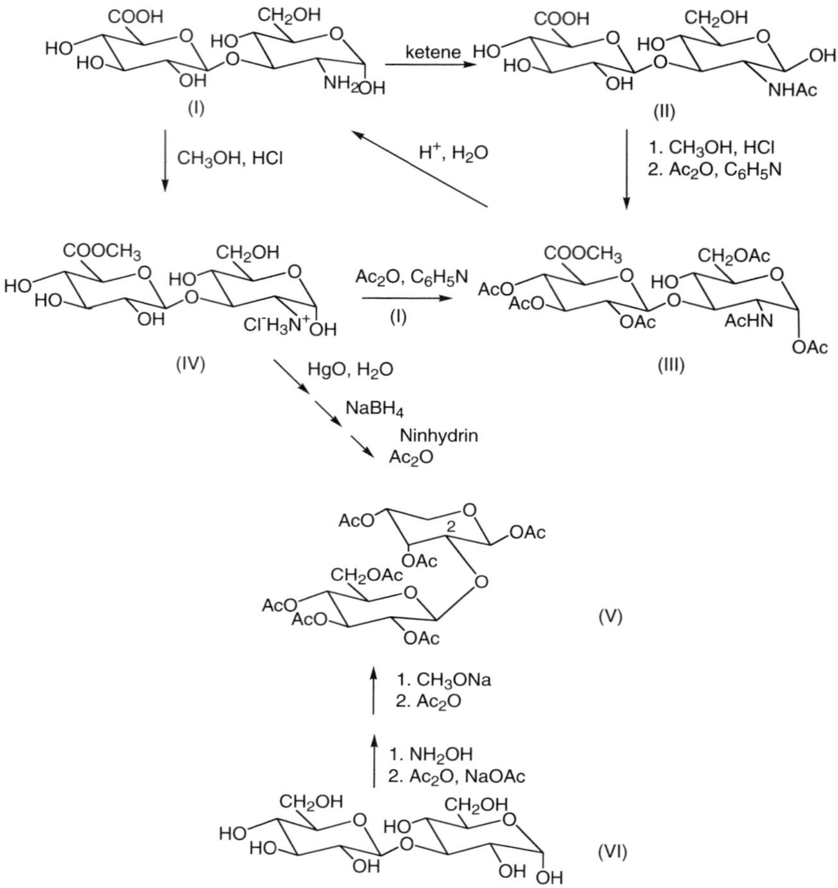

FIGURE 2.1
Hyaluronic acid structure investigation by Weissmann and Meyer using chemical reactions of hyalobiuronic acid (I) and a known disaccharide laminaribiose (VI). Hyaluronic acid disaccharide is N-acetylhyalobiuronic acid (II). Both (I) and (VI) produced the same disaccharide 2-O-beta-D-glucopyranosyl-D-arabinose heptaacetate (V).

with a melting point of 236°–238°C and a distinct optical rotation. An identical compound was also obtained through the condensation of two composing monosaccharides, 2,3,4,6-tetra-O-acetyl-α-D-glucosyl bromide and 2-acetamido-4, 6-O-benzylidene-2-deoxy-α-D-glucopyranoside, and through methanolysis of N-acetyl hyalobiuronic acid (Fig. 2.2).

The esters of HA disaccharides derived via alkylation and acylation are less polar derivatives than the original HA disaccharides, and they are soluble in and easily handled by organic solvents. Those were two early examples of the synthesis of water-insoluble derivatives of HA for the purpose of structural characterization. The very concept later evolved into using

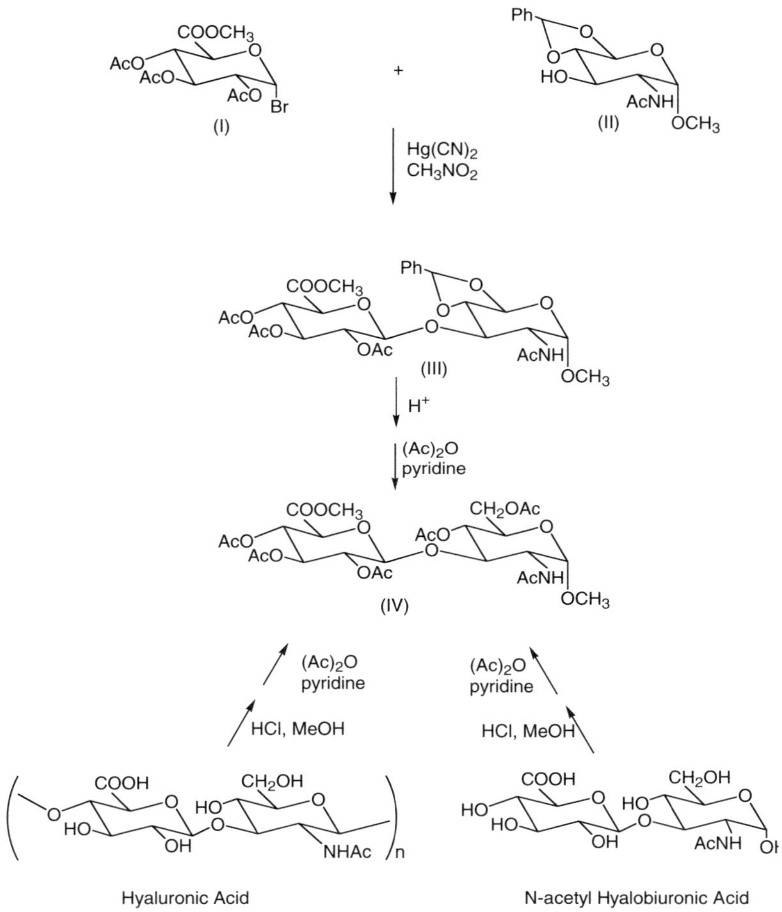

FIGURE 2.2
A fully acetylated and methylated disaccharide (IV) (MP = 236–238°C, optical rotation of +30°
in chloroform, c 0.68) was synthesized from hyaluronic acid, N-acetyl hyalobiuronic acid, and
the condensation product (III) of 2,3,4,6-tetra- O-acetyl-α-D-glucosyl bromide (I) and 2-aceta-
mido-4,6-O-benzylidene-2-deoxy-α-D-glucopyranoside (II).

water-insoluble derivatives of HA with hydrophobic side chains as potential
products for medical use (e.g., as biomaterials for drug delivery).

2.1.2 Silylation of HA and Its Use in Gas Liquid Chromatography

Lipophilic HA derivatives found their utility in gas liquid chromatography
(GLC) as early as the 1960s. Free carbohydrates have low volatility even
when reduced to smaller sizes. To make the GLC technique applicable to
carbohydrates such as HA, it is necessary to convert the sugar into its more

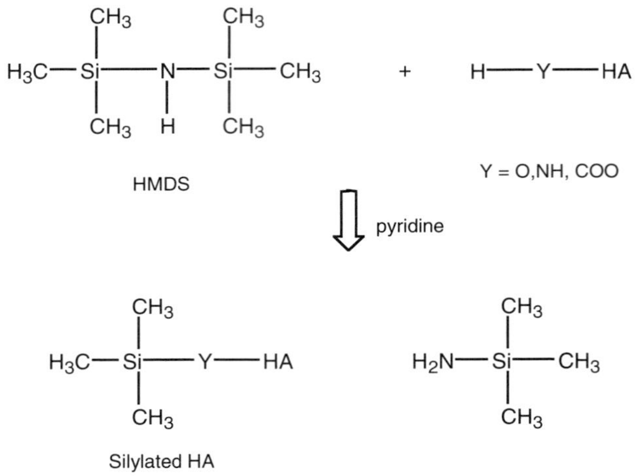

FIGURE 2.3
Silylation of hexosamine from the hydrolysis of hyaluronic acid. HMDS is the silylating agent hexamethyldisilazane.

lipophilic derivatives. Murphy et al.[4] prepared an ester derivative of HA via the reaction of acid-hydrolyzed HA and acetic anhydride. Another commonly used method is the silylation of the sugar molecule for the synthesis of its trimethylsilyl derivative. Radhakrishnamurthy et al.[5] used hexamethyldisilazane as silylating agent for hexosamine in GLC analysis. HA was first hydrolyzed in HCl to give hexosamine and then derivatized by hexamethyldisilazane to produce the volatile derivatives. The reactive functional groups (hydroxyl, carboxyl, and amino groups) of hexosamine were silylated (Fig. 2.3). The chromatograms of HA derivatives by the methods of both Murphy (acylation) and Radhakrishnamurthy (silylation) displayed distinct peaks from the chondroitin sulfate derivative by the same methods.

2.1.3 Peroxidation Reaction

Peroxidation normally occurs at the vicinal hydroxyl groups attached to the C2-C3 atoms of the uronic acid. As the result of this reaction, the C2-C3 bond is cleaved, the hydroxyl groups are oxidized, and aldehyde groups are formed. Montogomery and Nag found that the complete oxidation of HA by peroxide at room temperature took more than 10 days.[6] The effect of Donnan exclusion on the periodate as a result of the repulsion of a polyanionic charge of HA does not fully account for the slow reaction. D-glucuronic acid (HA or chondroitin 4 and 6-sulfate) has much slower oxidation than L-iduronic acid (dermatan sulfate), although both glycosaminoglycans have the same polyanionic charge density. Molecular models indicate that the slow reaction is related to the hindrance caused by the hydrogen bonding

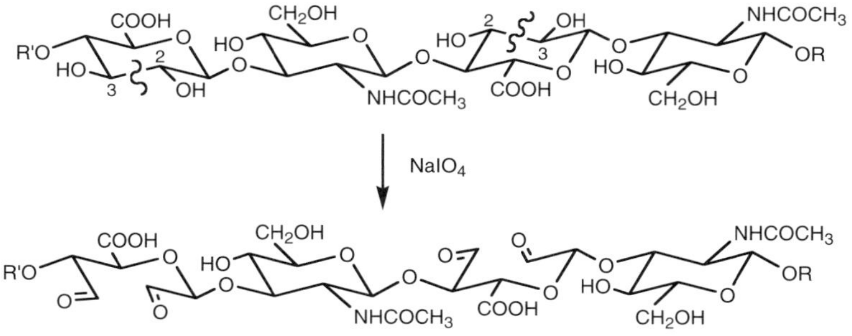

FIGURE 2.4
Oxidation of 2,3 vicinal hydroxyl groups of glucuronic acid forms two aldehyde groups. The low reaction rate is indicative of a hindered environment at the reaction site, probably caused by the hyaluronic acid helical structure.

of the carboxyl and C3 hydroxyl groups when the glycosaminoglycans are in certain conformation. This was actually part of the early evidence that indicated the existence of a hydrogen-bonded secondary structure for a HA molecule (Fig. 2.4).[7,8]

2.2 Chromogenic Reactions

Carbohydrates can be detected and quantified by various colorimetric methods. When treated with HCl, pentose decomposes to furfural (furan-2-aldehyde), and hexose decomposes to furfural derivative (5-hydroxymethyl furan-2-aldehyde). These furan-type compounds are chromogens that react with phenols, aromatic amines, and aliphatic amino acids, and in particular L-cysteine, to give colored products.[9]

One of the most commonly used colorimetric assays for HA is a carbazole assay introduced by Dische[10] in 1946. The colorimetric assay involved a two-step reaction, the first step being the dehydration of HA or other hexuronic acid with concentrated sulfuric acid. The reaction is different from Tollens' Reaction, in which carbohydrates are decomposed by HCl with the formation of furfural and carbon dioxide. In the reaction described by Dische, no gaseous product was released, and it was believed that the carbonyl group of the hexuronic was still intact. The second step was the reaction between the furfural derivative intermediate and carbazole, which yielded a red substance with maximum absorbance at 530 nm. The probable structures of the intermediate and the red substance are postulated as follows, without exact experimental evidence (Fig. 2.5).

The Elson-Morgan method to detect and measure the hexosamine was developed in 1933, before the discovery of HA. When treated with acetylacetone in

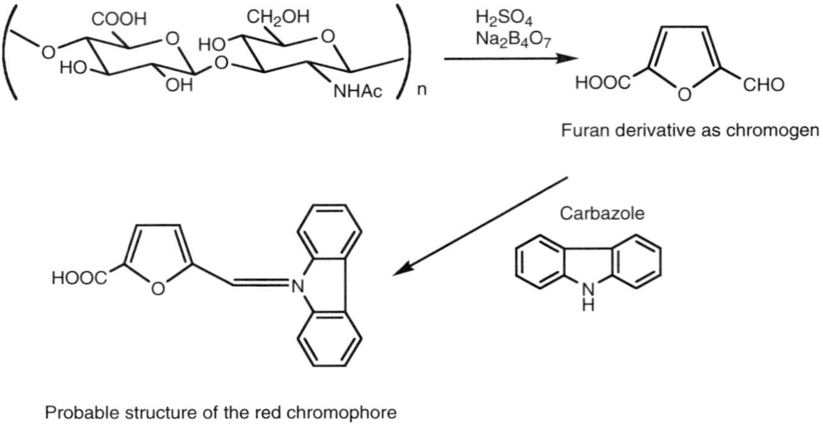

Furan derivative as chromogen

Carbazole

Probable structure of the red chromophore

FIGURE 2.5
Probable mechanism of reaction of hyaluronic acid with carbazole.

alkaline solution, hexosamine was converted to a pyrrole derivative, which gave a colored condensation product with DMAB (dimethylaminobenzaldehyde-Ehrlich reagent).[11,12] The Elson-Morgan assay had limited utility in the measurement of HA, because it requires the presence of amino group, as would be the case of hexosamine (Fig. 2.6).

The Morgan-Elson assay,[13] however, can be applied to N-acetyl hexosamine. It is similar to the Elson-Morgan method in that the intermediates of both reactions react with Ehrlich reagents to produce color. The main difference is that the Morgan-Elson reaction only needs a brief heating of the monosaccharide under mild alkali conditions, without requiring the participation of acetylacetone.

The development of the Morgan-Elson Assay is related to some findings from as early as 1901, when Erhlich observed, as mentioned by Aminoff et al.,[14] that mucins (containing acetylamino groups) gave rise to an intense purple color after heating with dilute alkali and the addition of DMAB. Morgan and Elson[13] suggested that the sugar reacted in enolic form under alkali conditions and formed glucoxazole or glucoxazoline by elimination of one water molecule (Fig. 2.7). White concluded in 1940 that amino sugar actually formed the oxazoline derivative under such conditions, and that the formed heterocyclic moiety of carbohydrate derivative then reacted with DMAB to produce a red color.[15]

The Morgan-Elson assay for N-acetylamino sugar was modified by Reissig et al.[16] Reissig method replaced the sodium carbonate with a concentrated borate buffer, improving the reproducibility of the assay even in the presence of protein or magnesium ions. It has since been widely used as one of the routine assays for HA.

The knowledge of the structures of the intermediates of the Morgan-Elson reaction continued to advance in the following decades. In the 1950s,

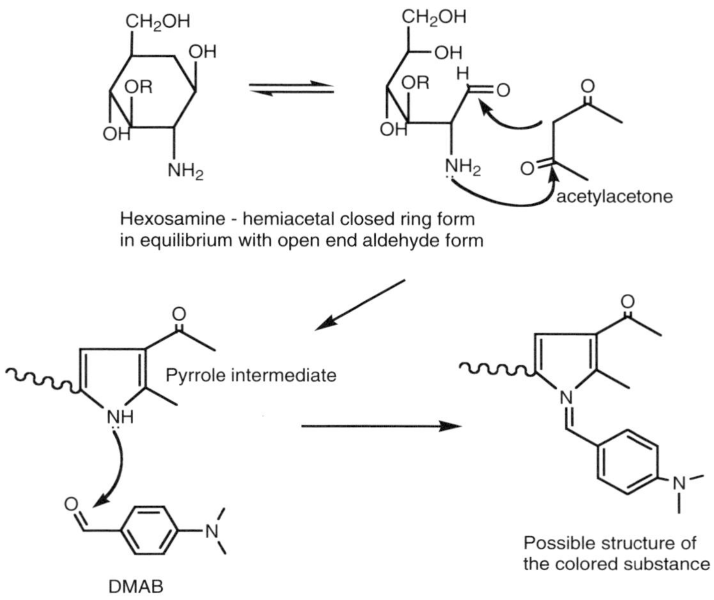

FIGURE 2.6
Elson Morgan reaction. Hexosamine at reducing end reacts with acetylacetone to form a pyrrole derivative that develops a color complex with dimethylaminobenzaldehyde (DMAB).

Kuhn et al. studied the structure of chromogen III using IR,[17] and in the 1970s, Beau et al. studied the structure of chromogen I using NMR and IR.[18] The authors concluded that the chromogen in the Morgan-Elson reaction was a furan-type of compound, not a heterocyclic intermediate, as previously described by Morgan and Elson. Benson also postulated that the N-acetyl group in the chromogens was intact during the Morgan-Elson reaction.[19] In the 1990s, Muckenschnabel et al.[20] published studies on the reaction products

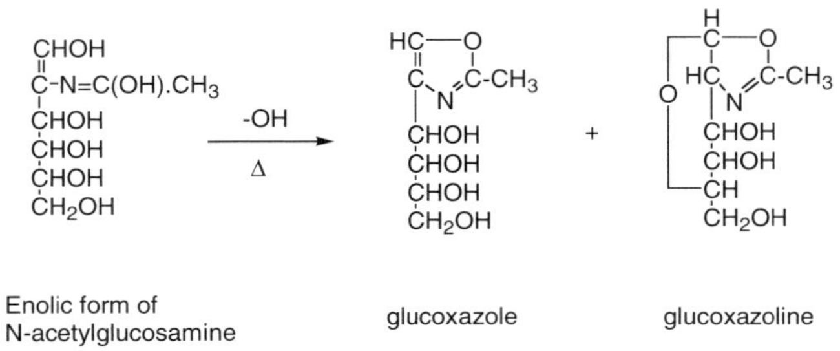

FIGURE 2.7
N-acetylglucosamine may be converted to glucoxazole and glucoxazoline, which further react with Ehrlich reagent to form color.

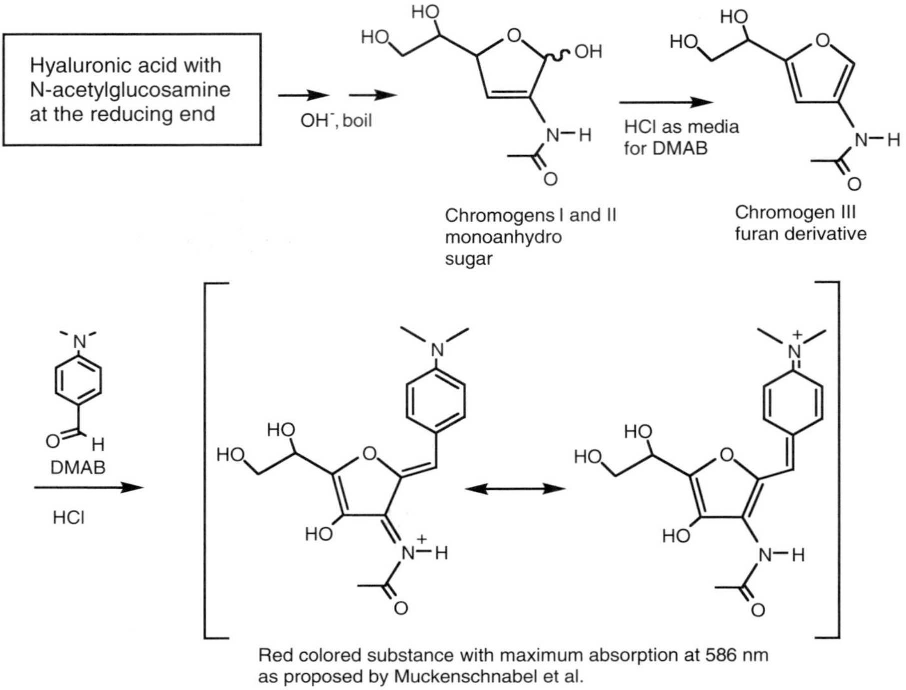

FIGURE 2.8
The furan-type chromogen intermediate formed by the reducing-end *N*-acylhexosamine and the formation of red color during Morgan Elson reaction — findings by Muckenschnabel et al.

of chromogen III and DMAB. On the basis of the data from HPLC, UV-Vis, and LC-MS, the authors proposed the structure of the red substance as mesomeric forms of N-protonated 3-acetylimino-2-(4-dimethylaminophenyl) methylidene-5-(1,2-dihydroxyethyl)furan (Fig. 2.8).

Although the detailed mechanisms of the chromogenic reactions (Dische, Elson-Morgan, Morgan-Elson, and Reissig) may not be exactly clear for general consensus, the common characteristics of those reactions appear to involve a furan derivative and an aldehyde.

2.3 Chemistry of HA Degradation

2.3.1 Alkaline Degradation

The α-hydrogen on the carbon atoms adjacent to carbonyl group is labile under alkaline conditions. This chemical property is by and large responsible for the degradation of HA when treated with strong alkali such as NaOH.

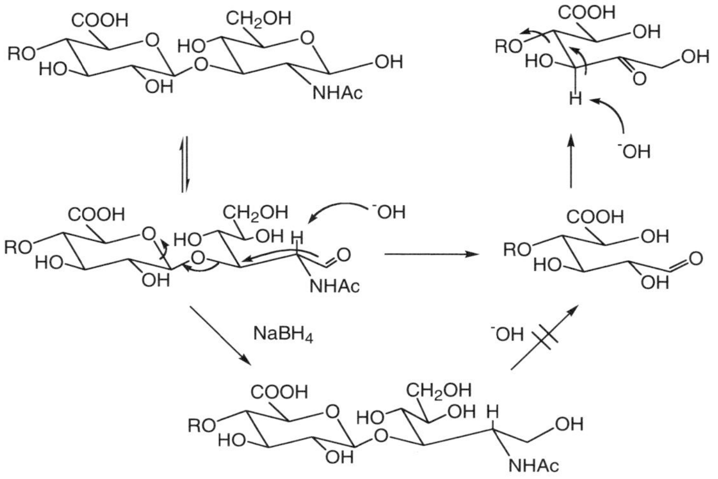

FIGURE 2.9
Progressive degradation of hyaluronic acid from the reducing end — "peeling reaction" — may be prevented by treatment with NaBH$_4$.

The N-acetylglucosamine of HA at the reducing end of the polymer chain exists as hemiacetal and aldehyde in equilibrium. The α-hydrogen of the aldehyde is reactive, and the neighboring glycosidic bond is prone to cleavage under strong basic conditions. The reaction would peel off the N-acetylglucosamine unit and expose the resultant glucuronate at the reducing end to further degradation. Such unintended progressive erosion of the polysaccharides from the reducing end can be prevented by the addition of NaBH$_4$. The reducing agent NaBH$_4$ reduces the aldehyde group to an alcohol group, rendering the neighboring hydrogen atom inactive, and thus preventing the stepwise degradation of the polysaccharide from the reducing end (Fig. 2.9).[21,22] However, for HA, the α-hydrogen of the aldehyde group at the reducing end is not the only labile hydrogen atom in the giant molecule. For every repeating disaccharide unit of HA, there is a labile H-atom at the 5 position of the glucuronic acid ring.

During the investigation of the colorimetric reactions of N-acetylglucosamines, Aminoff found that when HA was treated with hot dilute alkali or alkaline buffers, a substance was formed with maximum ultraviolet (UV) absorbance at 230 nm. This was observed by Kuo[23] in his study to deacetylate HA under basic conditions, similar to the deacetylation of colominic acid reported previously in the literature.[24,25] Thus, HA with a molecular weight (MW) of approximately two million was dissolved in sodium hydroxide containing 1 mg/ml NaBH$_4$, sealed in glass tubing, and heated at 105–110°C for 2 hours. Then the solution was dialyzed against flowing water for 2 days in a dialysis bag with a 12,000 MW cutoff. After lyophilization, the recovery of the treated HA turned out to be less than 2%. It is apparent that significant

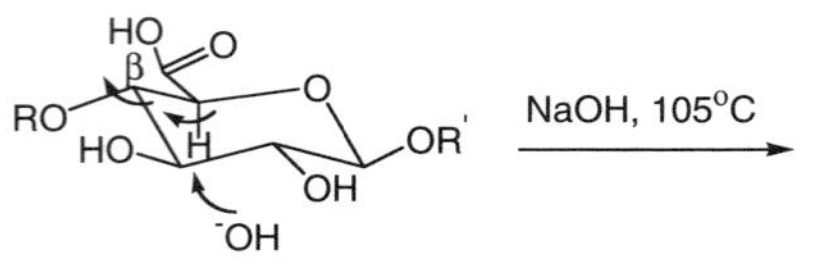

FIGURE 2.10
Hyaluronic acid undergoes β-elimination when treated with hot alkali and forms an α, β unsaturated carboxylate disaccharide that has a strong ultraviolet absorbance at 232 nm.

degradation occurred during the reaction, and the degradation probably did not just occur from the reducing end.

The UV spectrum of the degraded HA product showed a strong absorbance at 232 nm, which is characteristic of α, β unsaturated acid, as suggested by Linker et al.[26] Under the reaction conditions described, HA may have degraded at the 1→4 glycosidic linkages through β-elimination (Fig. 2.10).

2.3.2 Acidic Degradation

Glycosaminoglycans (GAGs) in general have considerable resistance to acid hydrolysis and its degradation under such conditions is rarely complete. Certain structural features of GAGs have a significant effect on the rate of acid hydrolysis. When the N-acetyl group is intact, the glycosidic bonds are readily hydrolyzed. As the deacetylation occurs simultaneously with the degradation of glycosides, the freed amino groups make the glycosaminide bonds more resistant to acid hydrolysis.[27] Moggridge and Neuberger studied the kinetics of acid hydrolysis of methylglucosaminide and explained that the inhibition of its hydrolysis may be a result of "the effect of adjacently positively charged amino-group in repelling hydrions from its immediate vicinity."[28]

In the study of the chemistry of HA degradation, the focus of interest has been on the nondestructive depolymerization reactions that produce discrete HA oligosaccharides. Inoue and Nagasawa prepared even-numbered HA oligosaccharides from high MW (2M Da) by acid hydrolysis.[29]

Marchessault and Rånby proposed an induction-stabilization theory of the effect of uronic acid on the neighboring glycosidic bonds.[30] BeMiller described such effect using carboxylcellulose as an example.[31] The presence of an electronegative carboxyl group at the C-5 position can exert inductive influence on the glycosidic oxygen atoms in two different ways. It facilitates the electron shift toward the glycosidic oxygen (making it more labile) that is closer to the carboxyl group, while opposing the electron shift to the glycosidic oxygen (making it more stable) that is situated farther toward the other side of the pyranoside ring (Fig. 2.11). If the same mechanism applies to HA, it could well explain why the even-numbered HA oligosaccharides with the glucuronic acid at the nonreducing ends become the predominant products when HA is treated with hydrochloric acid.[29]

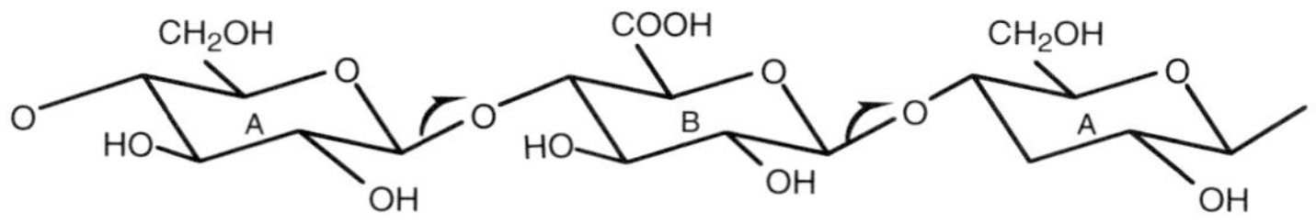

FIGURE 2.11
The carboxyl substituent at C6 of alternate disaccharide rings of cellulose may have an inductive effect on the electron shifts of neighboring glycosidic bonds. It favors electron shift from ring A to B but opposes the shift from ring B to A.

From a stereochemistry standpoint, the carboxyl group may serve as proton donor to the closest glycosidic oxygen — normally a first step toward depolymerization — and also serve as an intramolecular nucleophile to facilitate the breaking of the glycosidic bond, as also suggested by Roy and Timell and by Bochkov and Zaikov (Fig. 2.12).[32,33]

2.3.3   Oxidative Reductive Depolymerization

HA can undergo oxidative reductive depolymerization (ORD), in which both oxygen and reducing agents are involved. According to Pigman and Rizvi,[34] the reducing substances often used are ferrous or cuprous ions and ascorbic acid, as well as sulfhydryl (-SH)-containing organic compounds like cysteine and glutathione. Ferrous ions are the most effective reducing substances. Phosphate ions greatly speed up the reaction compared to the chloride ions. At 30°C and pH 7.3, the specific viscosity of HA could drop 90% in a time as short as 5 minutes. Synovial fluid showed similar degradations. The reaction is inhibited by glucose and completely repressed by a high concentration of alcohol.

Although the ORD reaction can degrade polysaccharides and nucleic acid in general,[35] evidence indicates that a glycosidic linkage near a hydroxyl group of a uronic acid polymer may be particularly susceptible.[34]

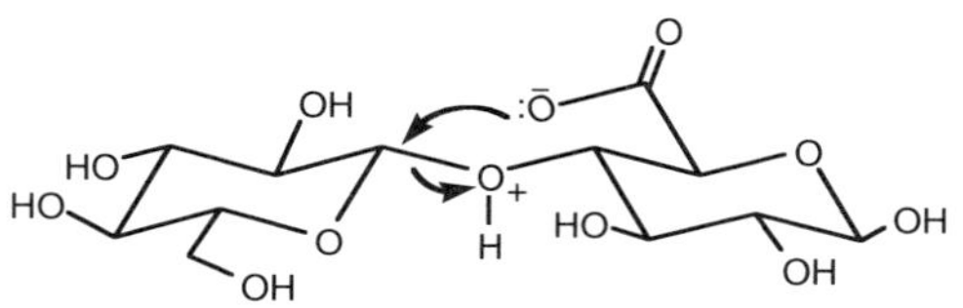

Cellobiouronic acid

FIGURE 2.12
A carboxylic group close to a glycosidic bond can be an effective intramolecular proton donor at pH ≥ pKa. The nucleophilic attack by the carboxylate may cause the decomposition of the oxonium ion, thus splitting the glycosidic bond.

The degree of degradation of HA by different ORD systems varied significantly. Swann treated HA with ascorbic acid alone, which reduced the MW of HA from over a million to 65,000 Da.[36] Cleland et al. reported the degradation of HA to its fragments with average MW of 24,000 under similar conditions.[37] Uchiyama et al. conducted an ORD reaction of HA by using Fe(II) in phosphate buffer and under an oxygen atmosphere. The starting HA had an average MW of approximately 400,000, and ORD reaction converted it to HA fragments within the range of 1000 to 10,000.[38] This confirms the conclusion by Pigman that Fe(II) is among the most effective reducing substances for ORD reactions.[34]

The oxidative effect of ascorbic acid in the degradation of HA is largely, if not exclusively, caused by the metal catalysis, which can exist as a contaminant in phosphate buffer and HA itself. At millimolar concentration, ferrous or cuprous ions cause depolymerization. At micromolar concentration, they are inactive alone but accelerate the rate of depolymerization by ascorbic acid. Ascorbic acid acts to regenerate ferrous ions by the following mechanism:[39]

$$\text{Ascorbic acid} + \text{Fe(III)} \rightarrow \text{dehydroascorbic acid} + \text{Fe(II)}$$

$$\text{HA} + \text{H}^+ + \text{Fe(II)} + \text{O}_2 \rightarrow \text{degraded HA} + \text{Fe(III)} + \text{H}_2\text{O}$$

Superoxide free radical $\text{O}_2^{-\bullet}$ is formed during the autoxidation reaction Fe(II) $\rightarrow$ Fe(III).[40,41] The free radical $\text{O}_2^{-\bullet}$ can also be formed by the action of certain oxidative enzymes such as xanthine oxidase. Accompanying the production of the superoxide is the production of hydrogen peroxide, as the latter is the reaction product of the former through dismutation.[42]

$$\text{O}_2^{-\bullet} + \text{O}_2^{-\bullet} + 2\text{H}^+ \rightarrow \text{O}_2 + \text{H}_2\text{O}_2$$

The superoxide radicals and the hydrogen peroxide further react to produce the hydroxyl radicals $\text{OH}^\bullet$

$$\text{O}_2^{-\bullet} + \text{H}_2\text{O}_2 \rightarrow \text{O}_2 + \text{OH}^- + \text{OH}^\bullet$$

McCord found that both $\text{O}_2^{-\bullet}$ and H_2O_2 were necessary for degradation of HA to occur. Superoxide dismutase is a scavenger for $\text{O}_2^{-\bullet}$. Catalase promotes the conversion of hydrogen peroxide to water and molecular oxygen. The addition of either superoxide dismutase or catalase protected HA against the free-radical degradation. McCord[42] concluded that the hydroxyl radicals $\text{OH}^\bullet$ are ultimately responsible for the degradation of HA in the ORD reaction.

Hydroxyl radicals can be created at the sites of inflammation where polymorphonuclear leukocytes may be activated by immune complexes and other opsonized material to produce oxygen-derived free radicals. These free radicals are capable of oxidizing tissue components to cause irreversible damage. This process can be inhibited by sufficient amount of superoxide dismutase.[42]

TABLE 2.1

Structures of Degradation Products from the Oxidative Reductive
Depolymerization Reaction of Hyaluronic Acid

Fragments from Reducing End	Fragments from Nonreducing End
4,5-unsaturated GlcUA($\beta1\rightarrow3$)N-acetyl-D-glucosaminic acid (21%) (see Structure B)	L-threo-tetro-dialdosyl-($1\rightarrow3$) GlcNAc (a tentative structure, 8%) (see Structure C)
4,5-unsaturated GlcUA($\beta1\rightarrow3$)GlcNAc ($\beta1\rightarrow3$)-D-arabo-pentauronic acid (24%) (see Structure A)	N-acetylhyalobiuronic acid (20%)
N-acetyl-D-glucosamine (51%)	N-acetyl-D-glucosamine (45%)

Uchiyama et al. investigated the mechanism of ORD reactions and the
structures of its degradation products in great detail.[38] HA was first degraded
to its fragments, with an average MW of 2600 Da by the actions of Fe(II) and
oxygen. Those fragments were then digested with chondroitinase AC-II. The
final reaction products as oligosaccharides and monosaccharides were separated by gel filtration and ion-exchange chromatography, and their structures were determined by NMR (proton and carbon-13) and fast–atom
bombardment mass spectrometry.[38] The results of the analysis of the final
degradation products are listed in Table 2.1, and some of their structures are
illustrated in Fig. 2.13. The study concluded, "the ORD reaction of hyaluronate proceeds essentially by random destruction of unit monosaccharides

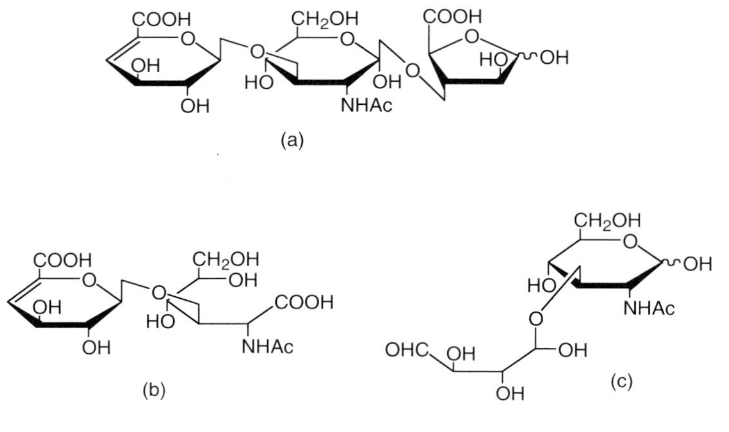

FIGURE 2.13
Structures of some degradation products from the oxidative reductive depolymerization reaction of hyaluronic acid: (a) 4,5-unsaturated GlcUA(β1-3)GlcNAc(β1-3)-D-arabo-pentauronic acid, (b) 4,5-unsaturated GlcUA(β 1-3)-N-acetyl-D-glucosaminic acid, and (c) L-threo-tetro-
dialdosyl- (β1-3)GlcNAc.

due to oxygen-derived free radicals, followed by secondary hydrolytic cleavage of the resulting unstable glycosidic substituents."

2.3.4 Degradation on Freeze Drying

Degradation of HA can occur during freeze drying. Wedlock et al. reported that free-radical scavengers such as Cl⁻, I⁻, alcohols, and sugars could inhibit this degradation.[43] It was concluded that freeze drying of NaHA caused mechanical stress and induced free carbohydrate radicals that depolymerized the polysaccharide. The produced hydroxyl radicals during freeze drying is attributable to carbohydrate radicals derived from C-C bond cleavage.[43]

Tokita et al. studied the degradation of HA in freeze-drying conditions by experimental methods such as electron spin resonance (ESR) and theoretical calculation of molecular dynamics. The researchers observed that freeze drying generates three times the amount of carbohydrate radicals from the acid form of HA as compared to the free radicals generated from the salt form of HA. Meanwhile, molecular dynamics calculation revealed three types of water molecules captured in the vicinity of HA molecules: (1) near the 1-4 glycoside pocket are the first type of water molecules that associate with the glycoside oxygen and an N-acetyl group through hydrogen bonding, as well as interacting with a carboxyl group through Coulombic force; (2) in the pocket near the 1-3 glycoside linkage are the second type of water molecules bonded to a glycoside bond, two hydroxyl groups, and an N-acetyl group; and (3) in the vicinity of carboxyl groups, water molecules are weakly captured through Coulombic force and observed only in the case of NaHA — not its acid form. Water molecules of types 1 and 2 are more tightly bond to the HA molecules and are named nonfreezing molecules, whereas water molecules of type 3 are more loosely associated with HA molecules, and so are named freezing water or free water. The authors concluded that water contributes to the stability of HA; the HA-H chain without water tends to be more mobile, and therefore tends to generate more free radicals, than the HA-Na chain without water; and HA-H also has higher reactivity toward the hydroxyl radicals as well.[44]

2.3.5 Radiolysis

It has long been known that HA solutions decrease viscosity when exposed to UV radiation.[45,46] A significant drop of viscosity of HA occurred even when the radiation dose was as low as 50 rads.[47]

Caputo reported in 1957 the depolymerization of HA by x-rays.[48] The irradiation of HA extracted from umbilical cord was carried out with soft x-rays (the wavelength range for x-rays is from about 10^{-8} m to about 10^{-11} m, and the corresponding frequency range is from about 3×10^{16} Hz to about 3×10^{19} Hz; hard x-rays are of higher frequency and are thus more energetic, whereas the lower-frequency x-rays are called soft x-rays). Sedimentation constant and electrophoretic patterns of the radiated HA were used to assess the degree of its depolymerization. The effect of irradiation (200×10^3 rads) on the degradation of HA was almost as dramatic as that of a hyaluronidase treatment. It was also found that the quantity of glucosamine as measured by Elson Morgan assay was directly proportional to the dose of x-ray. There-fore, it was concluded that the "ionizing radiations, like hyaluronidase, act on HA by opening the N-acetylglucosaminic bond."[48]

Balazs et al. studied the degradation of HA exposed to gamma irradiation (gamma radiation is high-energy photon emission resulting from natural radioactivity and is the most energetic form of electromagnetic radiation, with a very short wavelength [less than 10^{-10} m] and high frequency [greater than 10^{18} Hz]). Electron paramagnetic resonance was used to analyze the deg-radation products. The study concluded that "both glycosidic cleavage and attack of the C-5 hydrogen in the pyranose ring may take place to give the transient free radicals."[47] The exposure of HA to electron beam radiation with doses ranging from 1 to 4×10^6 rads not only reduced the molecular size of HA but also destroyed the chemical structures of hexosamine and hexuronic acid.[46]

The fact that HA is so susceptible to very low dose exposure to radiation has practically ruled out the use of radiation as a sterilization method for most HA-based viscoelastic products. According to *Remington's Pharmaceutical Sci-ences,* "radiation doses of 15 to 25 kGy are sufficient to kill the most resistant microorganism."[149] This is tantamount to the dose at mega rad (1 Mrad = 10 kGy) level, which is thousands times the dose that a typical HA polymer formulation can withstand.

2.3.6 Heat Degradation

It is well known that HA, especially when in the form of an aqueous solution, cannot withstand elevated temperatures for any significant amount of time. Many have observed the dramatic decrease of viscosity of HA solutions when subjected to conditions of autoclaving (e.g., 121°C for 12 minutes). The degradation product shows a strong UV absorbance at 232 nm, strongly indicating the breaking of glycosidic bonds through elimination reaction and the formation of α, β unsaturated carboxylate.

Lowry and Beavers investigated the thermal stability of NaHA in aqueous solution within a temperature range from 25° to 100°C.[49] NaHA with an average MW of 1.3 million Da was formulated into a 0.03% (wt/wt) solution at neutral pH. Samples sealed in ampoules were exposed to the set tempera-tures for up to 96 hours. Cannon-Fenske Routine Viscometers were used to measure capillary flow times (cft). The ratio (r) of the capillary flow times of the sample with certain exposure time (cft_t) over initial sample cft_0, that is $r = cft_t/cft_0$, is an indication of the degree of decrease of the viscosity of NaHA samples under the given condition.

The study indicated that at 25°C, the NaHA solution was stable, and a drop of viscosity of 10% would require many thousands of hours according to the extrapolation of the acquired data. As a contrast, a 10% drop of viscosity occurred in less than an hour at 90°C. The decline of viscosity over the time started to accelerate exponentially at 60°C. A particularly interesting observation of the study was that all decreases in the viscosity of NaHA samples started with an initial but transient increase of viscosity, as the samples were exposed to the heat. It was suggested that the higher temperature induces separation of HA chain segments that are hydrogen bonded together. Such an effect on viscosity is "tantamount to increasing the number and/or length of polymer molecules. The additive effect would then be overtaken and decline as simultaneous chain scission due to thermal degradation continues."[49]

2.3.7   Ultrasonic Depolymerization

Ultrasonication has been used as a conventional method of degradation to obtain polymers of lower MW. The generally accepted theory is that ultrasonic depolymerization proceeds by mechanical force. Propagation of acoustic energy causes rapid pressure variation to form small bubbles (cavitation) in the liquid. Subsequently, the bubbles collapse with large velocity gradients, and this collapse is responsible for the breakage of polymers.[50]

Miyazaki et al. investigated the effect of sonication intensity, temperature, HA concentration, coexisting cations, and ionic strength on the depolymerization of HA in solutions. Size exclusion chromatography with a low-angle laser light-scattering photometer was used to measure the size change of the HA molecules. The initial depolymerization rate k was found to increase with the sonication intensity, but the ultimate depolymerized MW (M_{lim}) generally converged to the same value. For example, continuous sonication with 55 W depolymerized the HA to a M_{lim} of approximately 0.1×10^6. However, the M_{lim} almost tripled in the presence of concentrated monovalent cation, such as in a solution of 7M LiCl.

Miyazaki et al. also found that the degradation product of HA by ultrasonification had a narrow distribution of MW (M_w/M_n slightly above 1). This is different from degradation of HA by heat, which broadened the MW distribution of HA by random scission. It was demonstrated that the low-MW samples with desired, narrow size distribution could be prepared by adjusting the sonication intensity or the constitution of the solution.[50]

2.4   Chemistry in Isotopic Labeling

The need for an isotopic labeling technique has been a significant impetus behind the research and development of many chemical processes of HA. The utility of such labeling is twofold. First, the labeled HA can play a role as a tracer in metabolic and pharmacokinetic studies. Second, the labeled

HA can serve as a probe for proteins and cell surface receptors that specifically bind to HA.

HA has ubiquitous presence in human and animal tissues, and HA is constantly metabolized in the body. There are obvious advantages of using radiolabeled HA in the study of HA metabolism.[51,52] Radiolabeled exogenous HA can be easily distinguished from the endogenous HA of the study subjects, and therefore it is not necessary to compare data about exogenous HA with an ever-changing baseline of endogenous HA. For the study of absorption, distribution, and elimination of exogenous HA, using radiolabeled HA is often the method of choice.

There are two major methodologies for radiolabeling HA. One is chemical modification, and the other is biosynthesis. Chemical methods involve chemical reactions of HA with radioactive reagents containing ^3H, ^{14}C, ^{125}I, ^{11}C, and so on. Sometimes it requires that the target HA be chemically modified before a radioactive moiety can be introduced. Biosynthesis uses cell culture or fermentation processes with radioactive precursors (^3H or ^{14}C) such as glucose, acetate, and so forth.[53,54] Biosynthetic techniques can provide radioactive HA structurally identical to the natural HA. However, the processes are laborious, and they do not provide much control of the MW of the HA process.

Several chemical methods have been reported in the literature.[55–62] Each method has limitations as well as values, depending on the particular intended applications.

2.4.1   Exchange Reactions

Tritium gas exchange reaction for tritium labeling[55] in theory should not alter the chemical structure of HA. However, it requires several days of exposure of HA solutions to tritium gas, during which serious degradation of HA can occur. Moreover, much of the tritium labeled to HA is labile, as it is associated with the hydroxyl groups of HA. It is not practical to completely remove the labile tritium, nor is the labeled HA stable under storage. This method appears to be applicable mainly for the making of radioactive HA oligosaccharides[56] with MW less than a few thousand dalton.

2.4.2   N-Acetylation

One of the chemical methods used to prepare a stable, tritium-labeled HA was to incorporate tritium as part of the methyl group in the N-acetyl group. As a first step, the N-acetylglucosamine moiety needs to be deacetylated. In addition to the alkaline deacetylation, mentioned earlier in this chapter, deacetylation of HA by hydrazinolysis was also experimented with.[23,57] Hook et al. adopted the method by Dmitriev, in which HA and other glycosaminoglycans were mixed with hydrazine, hydrazine hydrate, and hydrazine sulfate, sealed in tubes and heated at 105°C.[63] The macromolecular HA was from rooster comb source. No details were given about the effect of the

deacetylation reactions on the molecular size of HA. Kuo found that hydrazinolysis at 105°C significantly degraded polymer HA. The starting HA had
a MW over 1 million Da. After 3 hours, only 60% of the HA stayed above
12,000 Da, and after 6 hours 20% did so. The deacetylation was complete at
9 hours, as evidenced by the NMR data, when only 6% of the HA was
collected after dialysis with membrane of 12,000 MW cutoff.[23] The second
step for the labeling is the reacetylation of the free amino group just obtained
through deacetylation. In the study by Hook, ^{3}H acetic anhydride was used for
acetylation, and the reaction proceeded in the presence of $NaCO_3$ under low
temperature, kept low by ice (Fig. 2.14).

The reaction used to introduce tritium to HA by deacetylation and reacetylation, however, is not totally specific to the N-acetyl group. As a control
arm, the labeling of HA without prior N-deacetylation also resulted in the
incorporation of ^{3}H radioactivity, which amounted to 30% of the radioactivity
incorporated into a hydrazine-treated sample. This indicates that some of
the ^{3}H acetate groups may have bonded to the polysaccharide chain via O-
acetyl linkage. However, it was mentioned that the acetylation reactions
seem to have a minimum effect on the ionic property, size, and specific
interaction with chondroitin sulfate proteoglycan.[57]

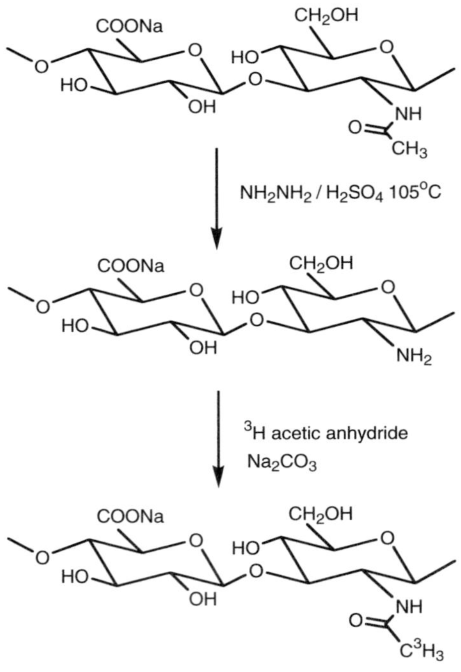

FIGURE 2.14
Deacetylation of hyaluronic acid by hydrazinolysis and reacetylation of the free amino group
to introduce tritiated acetyl group for radiolabeling.

2.4.3 Aldehyde Reactions

There are basically two types of aldehyde groups involved in the radiolabeling of HA. Aldehyde functional groups exist at the reducing end of HA in equilibrium with the closed ring hemiacetal form. Labeling only at the reducing end has the advantage of keeping the perturbation of HA structure to a minimum. The degree of labeling, however, is limited by the number of end groups and is inversely related to the MW of HA. As an alternative method, aldehyde functional groups can also be generated by peroxidation reactions in which vicinal hydroxyl groups are present. In the case of HA, the reaction occurs at the C2-C3 bond on the glucuronic acid ring, which cleaves the bond and yields two aldehyde groups. The peroxidation can thus produce a large quantity of reactive aldehyde groups for high-specificity radiolabeling without severing the polymer chains.

2.4.3.1   Formation of Cyanohydrin

The aldehyde group at the reducing end of carbohydrates can react with NaCN to form cyanohydrin. The method was named Fischer-Kiliani cyanohydrin synthesis[9] when used to add one carbon atom at a time to carbohydrate molecules. The reaction of HA hexasaccharide with ^{14}C potassium cyanide was reported by Robert et al.[64] Swann[58] used a similar method to measure the average number of reducing ends of HA with a MW of approximately 60,000 Da. It was found that the labeling method did not cause depolymerization, as the intrinsic viscosities of the HA before (192 ml/g) and after (187 ml/g) the reaction were basically unchanged (Fig. 2.15).

2.4.3.2   Reduction to Alcohol

As mentioned previously (see Fig. 2.9), the reaction of aldehyde at the reducing end with NaB^3H_4 can reduce terminal glucuronic acid to gulonic acid or reduce terminal N-acetylglucosamine to N-acetylglucosaminitol.[37] Orlando conducted a periodation reaction with HA (MW 500,000), and obtained HA with high specific radioactivity (0.15 mCi/mg). The reaction produces, for one repeating disaccharide unit, two aldehyde groups that can be reduced by NaB^3H_4 (Fig. 2.16).[62]

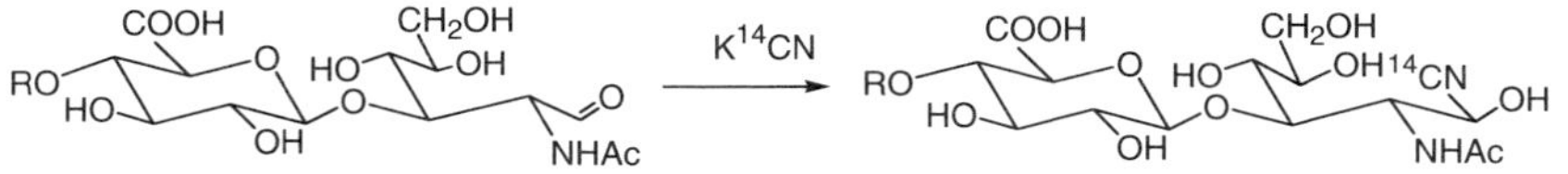

FIGURE 2.15
Reaction of KCN with aldehyde at the reducing end of hyaluronic acid forms cyanohydrin.

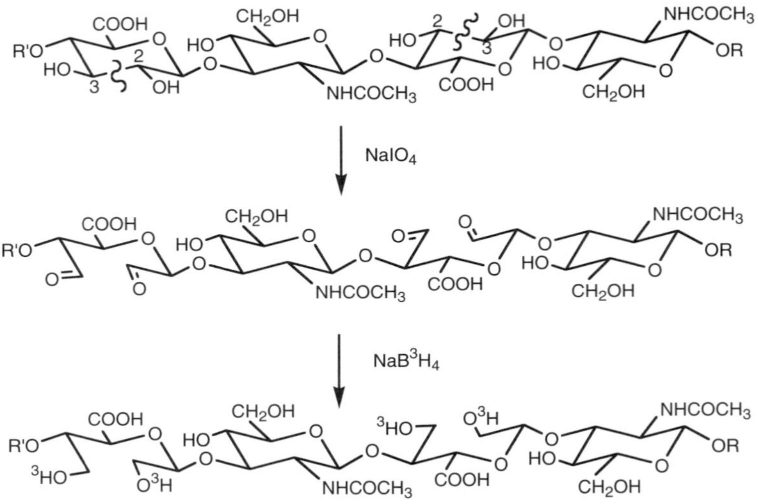

FIGURE 2.16
Partial oxidation of vicinal hydroxyl groups to form aldehyde, which can be reduced to alcohol by NaB^3H_4.

2.4.3.3 Alkylamine Bridge via Aldehyde

Raja et al. prepared an HA derivative containing a single hydroxyphenyl group at the reducing end, which can be radioiodinated.[65] The chemical modification of HA oligosaccharides (six to eight disaccharide units) consists of five major steps:

1. reduction of the terminal reducing sugar with sodium borohydride
2. sodium periodate oxidation of dihydro groups at the reducing end (GlcNAc) to generate a new aldehyde group at the end of the polymer chain
3. coupling of the aldehyde with diamine in the presence of sodium cyanoborohydride (reductive amination)
4. reaction between the amino derivative of HA and an active ester, Bolton-Hunter reagent
5. iodination of the hydroxyphenyl group. The product can be used as a radioactive probe for HA-binding molecules (Fig. 2.17).

2.4.3.4 Direct Conjugation with Tyramine

Orlando et al. also developed a method for radioiodination of HA at its reducing end. The HA used in the experiment had a MW of 170,000 Da. It is similar to Raja's method in that the HA molecules are iodinated at hydroxyphenyl group introduced to the reducing end of HA. Orlando's method, however, is straightforward, in that tyramine directly reacts with the aldehyde

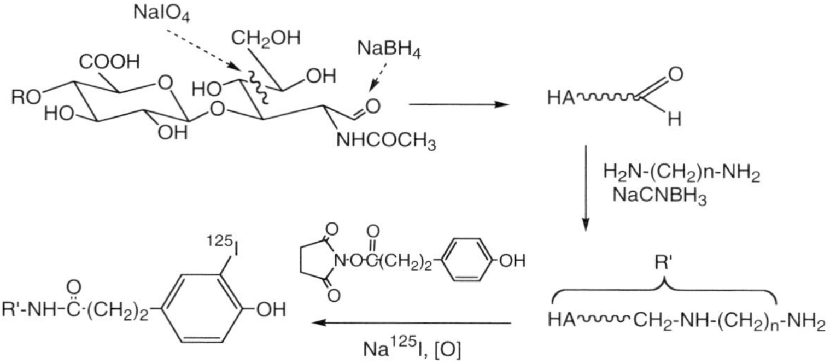

FIGURE 2.17
Preparation of alkylamine and ^{125}I-labeled hyaluronic acid oligosaccharide modified at the reducing end, a method by Raja et al.

at the reducing end of HA, and the formed Schiff base is treated with NaCNBH$_3$, followed by Na^{125}I labeling (Fig. 2.18).[59]

2.4.4 Hydroxyl Groups Activated by Cyanogen Bromide

The hydroxyl groups of HA may be activated by cyanogen bromide (CNBr) to form a cyclic reactive intermediate — imidocarbonate — which is susceptible to nucleophilic attack by amino groups. The formed N-substituted imidocarbonate can be converted to carbamate in final product (Fig. 2.19).[66]

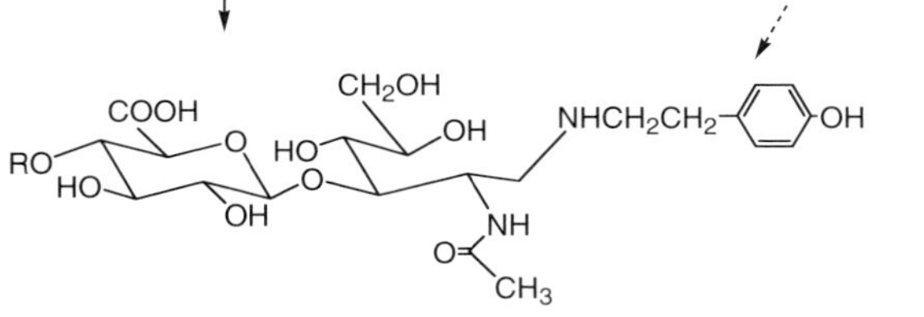

FIGURE 2.18
Radioiodination of tyramine- hyaluronic acid adduct. The amine nucleophile is in the hydroxyphenyl moiety. A method by Orlando et al.

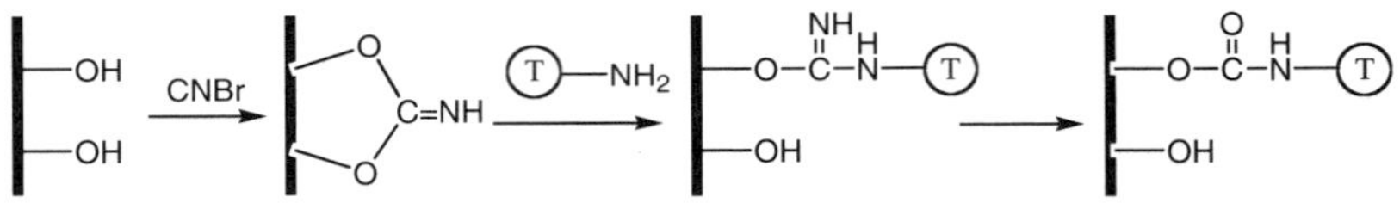

FIGURE 2.19
Immobilization of tyrosine or fluoresceinamine to hyaluronic acid with CNBr activation at vicinal diols. The reaction may involve the hydroxyl groups in two hyaluronic acid molecules.

Fluorescent labeling of HA, for example, can be obtained by the reaction of CNBr-activated HA and fluoresceinamine.[67] Radiolabeled HA[60] can be prepared by the reaction of tyrosine with CNBr-activated HA, followed by the treatment with [125]I. Cyanogen bromide can also be used alone to label HA. For instance, ^{11}C-labeled CNBr was synthesized to label HA for *in vivo* studies.[61] The synthesis would require special and expensive equipment such as a cyclotron. In addition, the radiolabeling of HA must be conducted at the application sites because of the extremely short half-life of [11]C (20.3 minutes).

2.5 Synthesis of Fluorescent HA

2.5.1 Isothiocyanatofluorescein

De Belder et al.[68] synthesized fluorescein derivatives of HA using isothiocyanatofluorescein, which targets mostly the primary hydroxyl group of HA. The reaction was conducted at 37°C overnight. The product showed an appreciable intrinsic viscosity drop from 1960 ml/g of the starting HA to 1465 ml/g of the labeled HA. The labeled fluorescent moiety was stable for over a month. The reaction probably occurred between the HA hydroxyl group and the isothiocyanate. The anticipated product was thiocarbamate, but the UV spectroscopy was inconclusive, due to overlapping absorption at 260 nm (Fig. 2.20).

2.5.2 Fluorescein Amine in Ugi Reaction

De Belder et al. employed the Ugi reaction to synthesize HA fluorescein derivatives from fluorescein amine.[68] The Ugi reaction involves four components: a carboxylic acid, an amine, an aldehyde, and an isonitrile (isocyanide).[69] The synthesis and the mechanism of making fluorescent HA using the Ugi reaction are illustrated in Fig. 2.21.

The reaction can be conducted under mild conditions in aqueous solutions. The formed amide linkage was stable after incubation at 37°C (pH 7.48) for

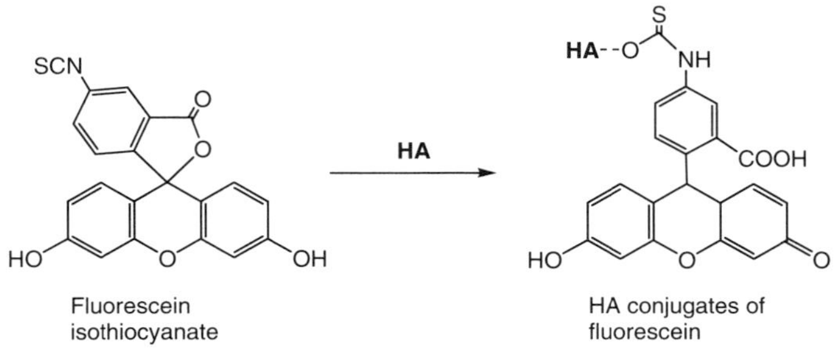

FIGURE 2.20
Fluorescein isothiocyanate reacts with hyaluronic acid to form thiocarbamate.

a month, as evidenced by T.L.C. (thin-layer chromatography). The depoly-merization during the reaction was not significant — a decrease of approx-imately 10% from 1960 to 1860 ml/g. For the two methods by de Belder, the degree of substitution ranged from 0.05 to 0.001. When isocyanide was used (Ugi reaction), the degree of substitution by fluorescein was 0.13%.

2.5.3 Fluorescein Amine in Carbodiimide Reaction

Ogamo et al.[70] employed carbodiimide (EDC) as the condensing agent in the reaction between HA and 5-aminofluorescein (FA). The reaction was believed to form an amide bond between the carboxyl function of HA and the amino group of FA, although the structure of the product was not verified.

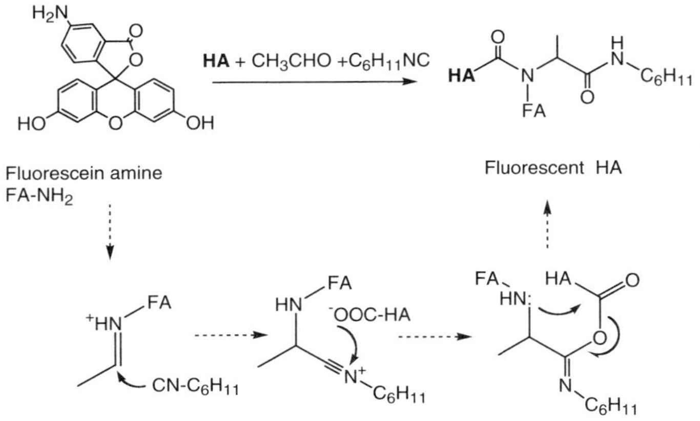

FIGURE 2.21
Fluorescein amine is conjugated to hyaluronic acid through a Ugi reaction — a four-component reaction of carboxylate, aldehyde, isocyanide, and amine.

HA from a rooster comb source with MW of 2 million was used. In the same study, the reactions of other GAGs were also run in parallel to the HA reaction. The fluoresceinated GAGs were then hydrolyzed and analyzed. One interesting observation was that HA had the lowest degree of substitution, at 0.0086, compared to chondroitin sulfate, dermatan sulfate, heparan sulfate and heparin, which as a group, showed a degree of substitution ranging from 0.015 to 0.040.

2.5.4   Synthesis of BODIPY-Labeled HA

BODIPY fluorophore was conjugated to HA by Luo et al.[71] The method is similar to the method used by Pouyani et al.[72] in terms of having the same reacting functional groups involved, hydrazide and N-hydroxysuccinimide (NHS). The difference is where the NHS function is located — with HA (Lou et al.) or the ligand (Poyani et al.) The reaction to make HA-BIODIPY conjugate is illustrated in Fig. 2.22. First, HA tetrabutylammonium (TBA) was formed by ion exchange, using the method described by della Valle.[73] The HA-TBA complex was soluble in DMSO. Then HA-TBA reacted with N-hydroxysuccinimido diphenyl phosphate to form HA-NHS, with a degree of substitution between 50 and 80 mol%. Finally, the HA-NHS reacted with BODIPY-FL hydrazide to afford the HA-BODIPY conjugate. The unreacted active ester was consumed by the addition of 2-aminoethanol (Fig. 2.22).

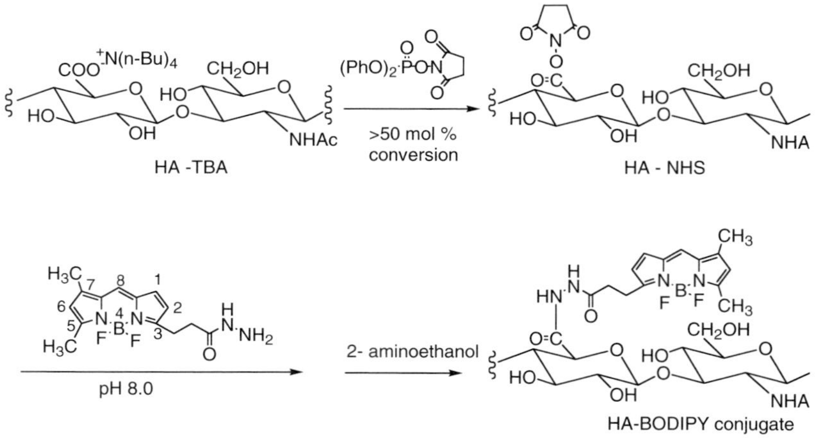

FIGURE 2.22
Synthesis of hyaluronic acid-BODIPY conjugate as fluorescent probe. TBA: tetrabutylammonium salt. NHS: *N*-hydroxysuccinimide. BODIPY: 4,4-difluoro-4-bora-3a,4a-diaza-*s*-indacene.

2.6 Biotinylated HA

Biotinylated HA is quite useful as a probe for detection of binding proteins in cells and tissues. The synthesis of biotinylated HA as its carboxyl function derivatives has been reported.[72,74–76]

Yu[74] and Yang[75] employed commercially available biotin-ε-amino-caproyl hydrazide. The first step was to form an O-acylurea of HA by reaction with EDC. The intermediate then reacted with the hydrazide to form HA-biotinylated HA (Fig. 2.23).

In the method by Pouyani et al.,[72] the hydrazide function is attached to HA instead. Hydrazide functionalized HA was synthesized by using excess adipic dihydrazide. The hydrazido-HA then reacted with sulfo-NHS-biotin to form the HA–biotin conjugate (Fig. 2.24). The method by Pouyani has the flexibility of designing the spacer arm by selecting the length of the dihydrazide. Two hydrazido linkages exist in the biotinylated HA using this method. NHS ester is one of the most effective chemical activation methods for making reactive acylating agents. It has been widely used since its introduction as a reactive crosslinking agent by Bragg and Hou,[77] as well as by Lomant and Fairbanks,[78] in the 1970s.

Frost and Stern also used biotin-hydrazide to prepare biotinylated HA in the presence of sulfo-NHS and EDC.[76] It was suggested that the activated NHS ester of HA, formed *in situ* by the actions of NHS and EDC, reacted

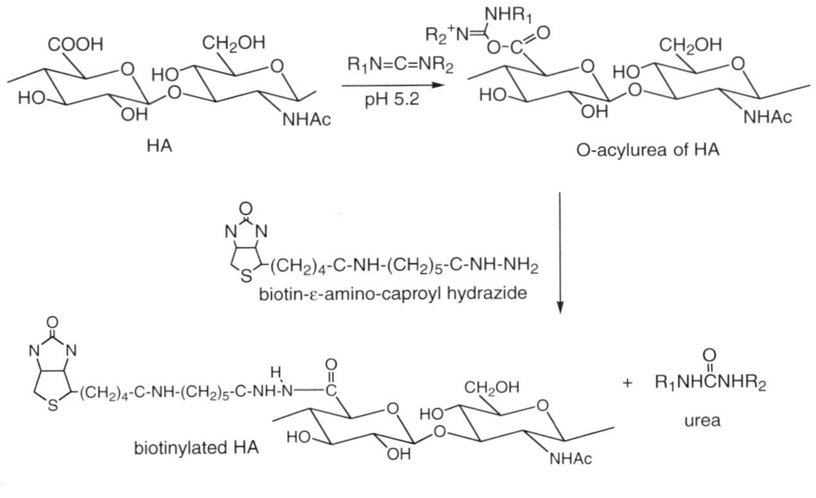

FIGURE 2.23
Synthesis of biotinylated hyaluronic acid by the reaction of hyaluronic acid and biotin-ε-amino-caproyl hydrazide in the presence of carbodiimide.

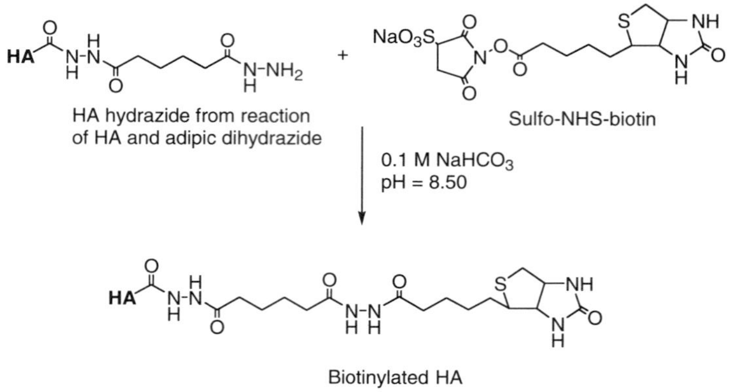

FIGURE 2.24
Synthesis of biotinylated hyaluronic acid by reaction of hyaluronic acid hydrazide and sulfo-NHS-biotin.

immediately with the biotin-hydrazide to form biotinylated HA.[79] The remaining carboxyl functions of biotinylated HA were activated again by NHS and EDC and were then immobilized to a microtiter plate, presumably by reacting with the amino groups on the plate to form an amide bond. This was done as part of the effort to set up a microtiter assay for hyaluronidase activity, which involved further interaction with avidin-peroxidase reaction in an ELISA assay.[76]

2.7   HA Derivatives with Various Side Chains

Chemical modification of HA by introducing side chains of various features (e.g., hydrophobic or cationic) may alter the physical and chemical properties of HA. As compared to native HA, the HA derivatives with side chains may be less water soluble, have increased residence time for *in vivo* application, and have enhanced interaction with other biologically active molecules such as drugs.

2.7.1   Acyl-Urea Derivatives

Kuo et al. studied the reactions between the carboxyl groups of HA and various carbodiimides in the presence and absence of primary amines at pH 4.5.[80] The reactions between HA and carbodiimides proceeded rapidly at room temperature. However, the final product was not amide, as claimed in some patent literature,[81] but N-acylurea or O-acylurea. Furthermore, to utilize the efficiency of the HA-CDI reaction, a series of designed carbodiimides

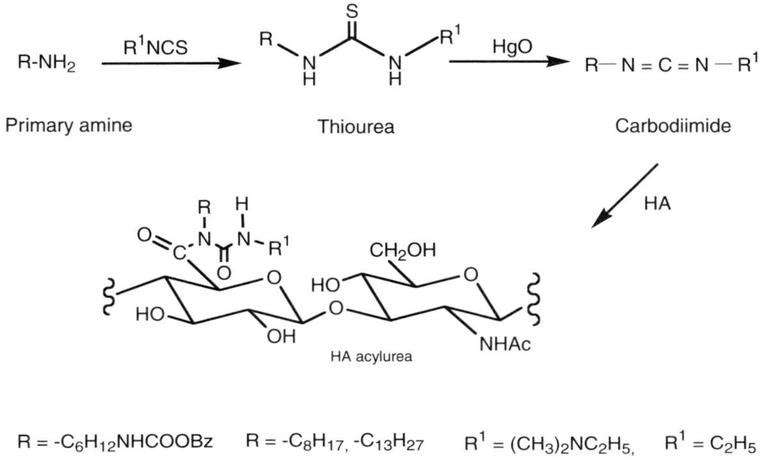

FIGURE 2.25
Synthesis of designed carbodiimides. The reaction products of hyaluronic acid and carbodiimides are hyaluronic acid–acylureas. The terms *R* and *R*[1] may be interchangable and represent hydophobic, aromatic, or cationic side chains.

and their corresponding HA acyl-urea derivatives were synthesized. The carbodiimides were synthesized in a two-step reaction: the reaction of primary amine and isothiocyanate to form thiourea, and the reaction of thiourea and HgO to produce carbodiimide (Fig. 2.25).

The HA acylurea derivative is one of the components of a commercially available HA-containing product for prevention of postsurgical adhesion (Seprafilm).[82,83] The other polymer in Seprafilm is carboxymethylcellulose (CMC), which presumably reacts with carbodiimide as HA does. The acylurea derivative has a certain amphoteric property resulting from the partially cationic nature of the carbodiimide EDC and the urea. In contrast, the unreacted carboxyl group on HA and CMC remain anionic. This cation and anion interaction renders the HA derivative less water soluble than the unmodified HA.

2.7.2 Water-Insoluble HA Esters

HA esters can be prepared by the method described by della Valle et al.[73] Briefly, HA was first converted to its quaternary ammonium salt by eluting HA solution through an ion exchange column of sulfonic resin in tetrabutylammonium form. The HA–quaternary ammonium complex dissolved in DMSO and reacted with esterifying agent such as alkyl or benzyl halides. The reaction presumably proceeded through nucleophilic substitution, with the hydroxyl oxygen in the carboxyl group of HA acting as nucleophile and the hydrocarbyl halogen in the halides as the leaving group. By selecting halides of various kinds, such as aliphatic or aromatic, and by controlling

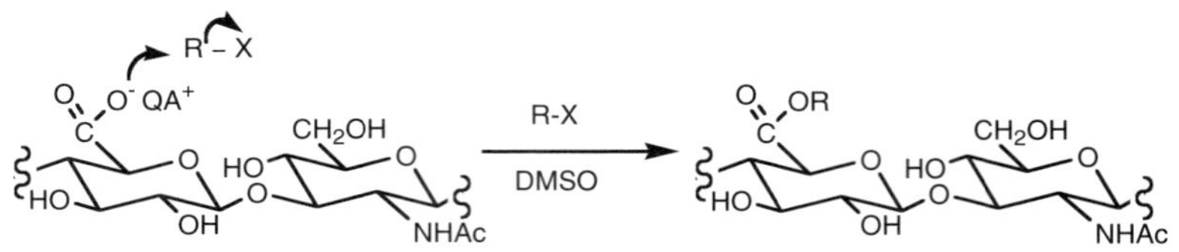

FIGURE 2.26
Synthesis of hyaluronic acid esters by reaction of the quaternary ammonium salt of hyaluronic acid with hydrocarbyl halide.

the degree of substitution, a series of HA esters (HYAFF) with different degrees of solubility have been synthesized by Fidia Pharmaceutical Corporation, using this method (Fig. 2.26).

2.8   HA–Drug Conjugation

Many of the chemical methodologies mentioned above have also been used to chemically link drugs to HA. It is hoped that drug–HA conjugates can improve the pharmacological properties of the targeted drugs.[84]

2.8.1   Esterification of HA

The esterification method by Della Valle et al. described above[73] was used to chemically link steroid and HA. In the esterification reaction, steroid serves as alcohol for its primary hydroxyl group at the 21-position, and HA serves as the carboxylate. It was claimed, for example, that 21-bromo-hydrocortisone reacted with HA-TBA complex, and that the degree of esterification amounted to 20%.

2.8.2   Hydrazide–NHS Ester Reaction

Direct conjugations of drugs to HA to attain significant drug load are rarely successful because of the insufficient reactivity between HA and most of the drugs under mild conditions. A practical approach is to convert HA and drugs into their more reactive derivatives. As preparative steps that can be taken before HA drug conjugation, HA can be attached with hydrazido functionality, and drugs can be attached with functionality of NHS activated ester, or vice versa.

   Thus, hydrocortisone-hemisuccinate was converted to its NHS-activated ester, and HA was attached with a hydrazido function by reacting with, for example, adipic dihydrazide in the presence of EDC. The covalent bonding

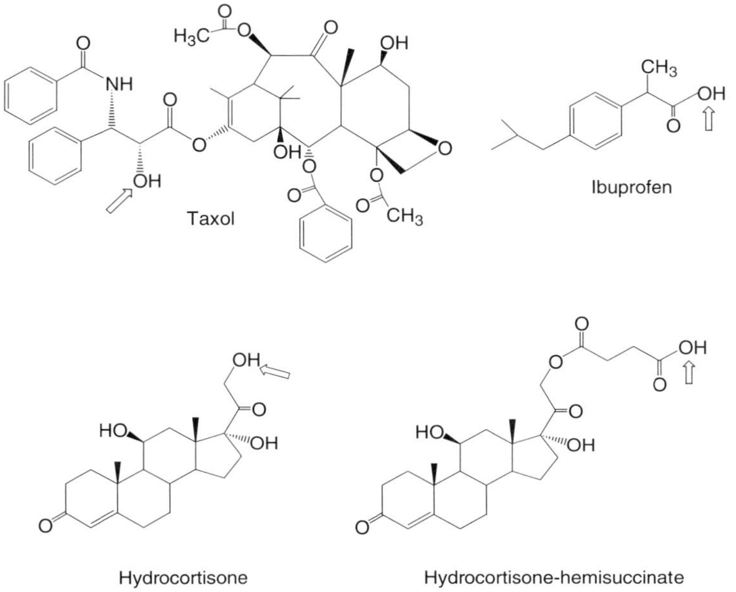

FIGURE 2.27
Taxol, ibuprofen, and hydrocortisone and its hemisuccinate — drugs covalently bonded to hyaluronic acid carboxyl group through the hydrazide-carboxylate reaction.

of the two forms the HA–hydrocortisone complex. The HA–ibuprofen conjugate was synthesized using the same method.[85]

Taxol ($C_{47}H_{51}NO_{14}$, MW 853.92) a natural compound from the bark of the Pacific yew tree and a potent cytotoxic agent with a fairly complicated structure, was covalently bonded to HA by Luo et al. Again this worked as an NHS ester and hydrazide reaction. The hydroxyl group of Taxol first reacted with succinic anhydride to form Taxol-2-hemisuccinate, which then reacted with N-hydroxysuccinimide diphenyl phosphate to form the Taxol-NHS ester. The counterpart was HA–hydrazide, a reaction product of adipic dihydrazide and low-MW HA obtained through partial degradation of HA by hyaluronidase (Fig. 2.27).[86,87]

2.9 Chemical Crosslinking of HA

One of the prominent characteristics often associated with a high-MW HA product is its extraordinary viscoelasticity and gel-like properties at relatively low HA concentration. For instance, an HA formulation with a MW of 2 million Da at a concentration as low as 1% w/v in a saline solution can be too viscous to fall off an inverted, small beaker by gravity only. Many HA researchers have aspired to improve on such unique properties of HA both for its intellectual worth and for its potential value as a novel biomaterial as well.

2.9.1   Divinylsulfone

Crosslinking of high-MW HA (2.5 million Da) by formaldehyde[88] and divinylsulfone (DVS)[89] can produce highly swollen gels. The methods represent the technology used in the manufacturing of hylan products, now being marketed under the trade name Synvisc to treat patients with osteoarthritis of the knee. Hylan is a generic term referring to HA crosslinked via its hydroxyl groups, leaving the carboxylate and acetamido groups unreacted.[90] The retention of the carboxylate group was considered critical to maintaining the polyanionic character of HA, and therefore it is important to preserve its natural physiochemical and biological properties. Hylan A is synthesized *in situ* by treating HA-rich tissues with formaldehyde before extraction. Hylan B is formed by the reaction between HA (or hylan A) and DVS under mild alkaline conditions. The chemical reaction between DVS and the hydroxyl groups of HA forms sulfonyl bis-ethyl linkages. The crosslinking produces an infinite HA network that is no longer water soluble (Fig. 2.28).[90,91]

DVS has been used to couple nucleophilic ligands to polysaccharides such as agarose. DVS reacts with hydroxyl groups of agarose under alkaline conditions (pH 11), and the products tend to be unstable above pH 9. DVS reacts with thiols and amino groups at lower pHs and at somewhat higher rate, and the products tend to be unstable above pH 8.0.[66,92,93]

2.9.2   Biscarbodiimide

Kuo et al. synthesized biscarbodiimides for the purpose of crosslinking HA.[23,80] The same synthetic route via thiourea was followed as the case for preparing the monocarbodiimides mentioned earlier. Thus 1,4-phenylenediamine and 1,6-diaminohexane were used as the starting material to prepare the corresponding aromatic and aliphatic biscarbodiimides. The aromatic biscarbodiimide appeared to be more reactive, as evidenced by its higher percentage of proton uptake compared to the aliphatic biscarbodiimide.

$$HA\text{-}CH_2\text{-}OH \longrightarrow$$

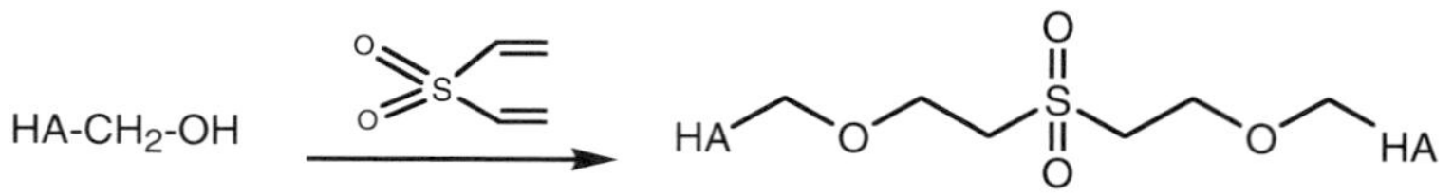

FIGURE 2.28
Crosslinking of hyaluronic acid by divinylsulfone in alkaline media.

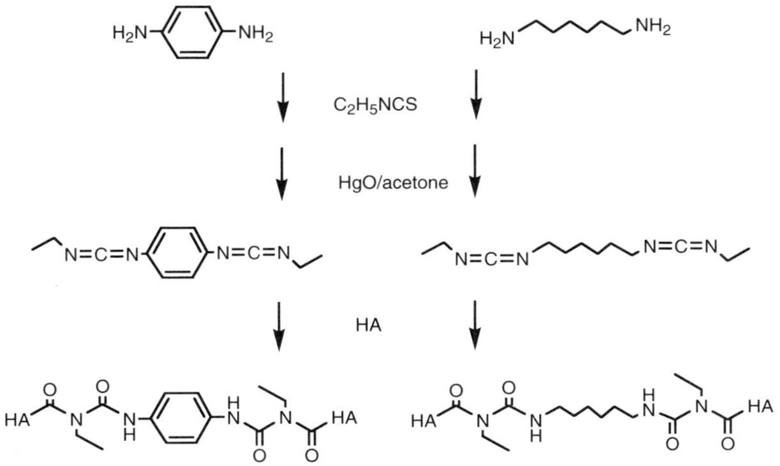

FIGURE 2.29
Synthesis of aromatic and aliphatic biscarbodiimides and their crosslinking reactions with
hyaluronic acid to form a hydrogel.

The reaction of HA and 18% equivalent of the phenylene biscarbodiimide
formed a highly swollen gel in saline solution (Fig. 2.29).

2.9.3   Crosslinking via Hydrazide

2.9.3.1   *Bis-NHS Crosslinkers*

Pouyani and Prestwich[85] used the reactions between hydrazide and an active
ester (e.g., NHS) to make crosslinked HA. The hydrazide function is introduced to HA by its reaction with dihydrazide in the presence of EDC. Then
commercially available sulfosuccinimidyl compounds are applied as
crosslinking agents. The chemistry is similar to the making of biotinylated
HA as illustrated in Fig. 2.24.

2.9.3.2   *Di- and Polyvalent-Hydrazide*

Vercruysse et al.[94] synthesized di-, tri-, tetra-, penta-, and hexa-hydrazides
that crosslink HA in the presence of carbodiimide. The hydrazide crosslinkers were prepared by the reactions between hydrazine and carboxyl acid
esters (Curtius Reaction).[95,96] The esters, in turn, can be synthesized by standard esterification of carboxylic acids and alcohols. For the synthesis of
polyvalent hydrazides, polyamines were used as the structural components
and the starting material of the crosslinkers. The polyamines reacted with
methyl acrylate to form the esters needed for the Curtius Reaction with
hydrazine to form the hydrazide crosslinkers (Fig. 2.30).[97]

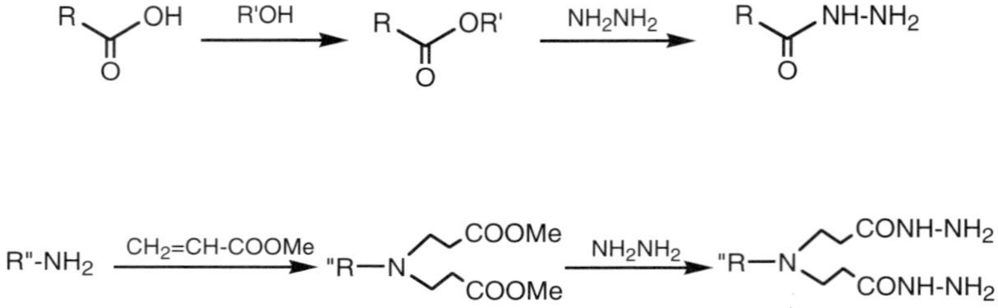

FIGURE 2.30
Hydrazide crosslinkers were synthesized by the reaction between hydrazine and carboxylic ester–Curtius reaction. The esters were prepared by standard acid–alcohol condensation or through the reaction of amine and acrylic ester.

The chemical reaction of HA with hydrazide crosslinkers proceeded within a controlled pH range (3.5–4.7). The reaction consumed acid, and HCl was added to maintain the pH level while rapid heterogeneous gelation occurred. In the presence of amine buffer salts, the formed gels were more homogeneous. Interestingly, the stability of the hydrogels in the presence of hyaluronidase was more related to the structural features of the crosslinkers than the degree of crosslinking. For instance, the gel made of a more hydrophobic crosslinker was more resistant to hyaluronidase degradation than was the gel made of less hydrophobic crosslinker.[94]

2.9.3.3   Bis-Aldehyde

The hydrazide derivative of HA is also reactive toward aldehyde. Thus, HA adipic dihydrazide can be crosslinked rapidly by polyethyleneglycol propiondialdehyde (PEG-diald) under acidic and neutral conditions to form a hydrazone linkage.[98] The crosslinked film has the property of fast gelling and fast swelling (Fig. 2.31).

2.9.4   Diepoxide

Diepoxides can react with the 6-hydroxyl group of HA under alkaline conditions to form ether linkages, whereas under acidic conditions, the reaction product of diepoxides and HA is an ester.[99] Diepoxides were used to crosslink agarose in NaOH solutions. Laurent used 1,2,3,4-diepoxybutane to crosslink HA with a MW of 1.5 million Da. To form a gel with a high swelling capacity, the lowest HA concentration needed for the reaction was 5% w/v.[100] Other examples of diepoxides are 1,4-butanediol diglycidyl ether (BDDE) and 1,2-ethanediol diglycidyl ether (EDDE) (Fig. 2.32).

Q-Med has been issued a HA process patent,[101] and its key technology appears to be a controlled process by which HA was crosslinked by diepoxide such as BDDE. The HA used was from a nonanimal source, and the

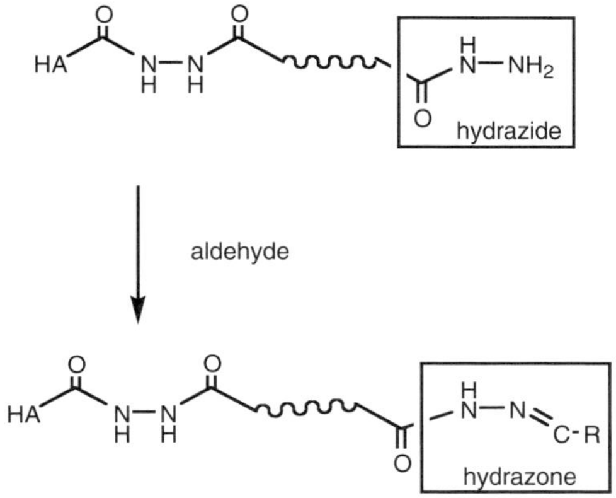

FIGURE 2.31
Dihydrazide and hyaluronic acid react to form hydrazido–hyaluronic acid with one pendent hydrazide function. The free hydrazide groups then react with aldehyde to form a hydrazone linkage.

crosslinked HA produced according to the technology is named NASHA (nonanimal stabilized HA), ("NASHA is the technology on which all Q-Med's HA products are based," Dagens Industri, 2000-07-12, Q-Med applies for DEFLUX approval in United States).

2.9.5 Self Crosslinking

Carboxyl function of HA generally requires activation to become reactive, under mild conditions, toward hydroxyl, amino, and other nucleophiles.

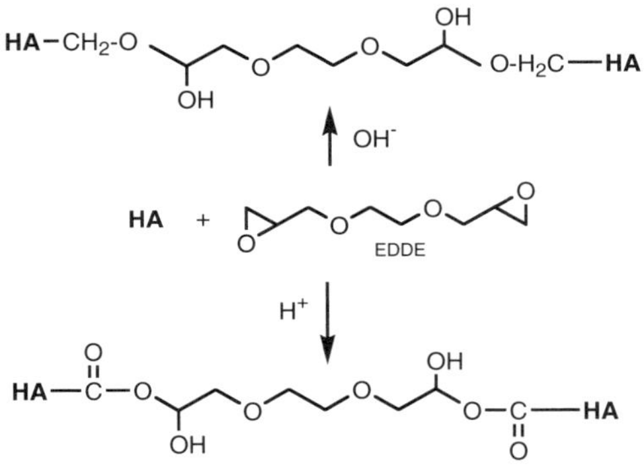

FIGURE 2.32
Diepoxide crosslinking of hyaluronic acid under alkaline and acidic conditions.

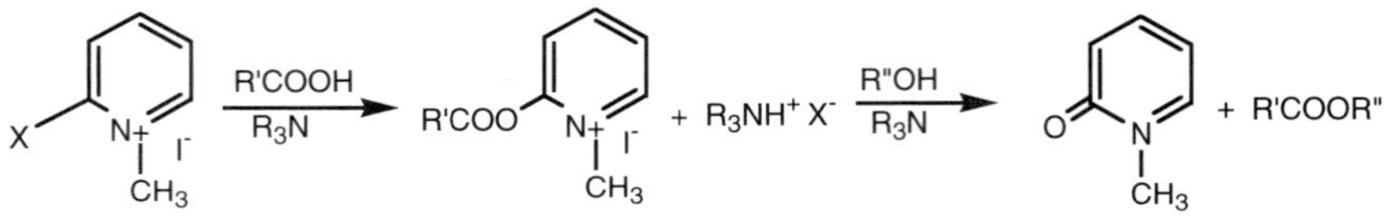

FIGURE 2.33
Self crosslinking of hyaluronic acid via internal esterification using 2-chloro-1-methyl pyridinum
iodide or 2-bromo-1-methylpyridinium iodide (Mukaiyama agent) as coupling agent.

It is known that carboxylic ester in general can be obtained in good yield by
the condensation of equimolar proportions of a carboxylic acid and an alcohol in the presence of a carbodiimide at room temperature.[102] HA activated
by NHS (N-hydroxysuccinimide) became reactive toward hydrazido
BODIPY (compound) (see Section 2.5).

   Another useful activating agent for HA is 2-chloro-1-methyl pyridinium
iodide (Mukaiyama reagent). The halogen atom at the 2-position is labile
and readily replaced by the nucleophilic carboxylate ion. The active acyl
intermediate then can react with the hydroxyl group to form a stable ester
(Fig. 2.33).[103,104]

   Della Valle[105] used the Mukaiyama reagent for the internal esterification
of HA. The intra- or intermolecular condensation of the carboxyl and
hydroxyl groups of HA was also called a self or auto crosslinking reaction.
HA was first converted to quaternary ammonium salt and dissolved in
organic solvent such as DMSO, where the self crosslinking proceeded
through the action of the Mukaiyama reagent and tertiary amine.

2.9.6   Multivalent Cations

Schmut et al.[106] were the first to produce stable gels from HA and metal ions
(Cu^{+2}) at physiological pH (6.0–6.8). (HA was known to form a viscoelastic
gel at pH 2.5 as a result of the favorable H bonding formed by a balance of
the ionic and nonionic carboxylic group[107].) The stable, transparent blue gel
was formed with a mixture of 2 mg/ml HA and 1 mg/ml $CuSO_4$. The critical
factor for gelation is the MW of HA. The gel formed when HA MW was
200,000 Da or more but failed to do so when the HA MW was at about
5000 Da. A carboxyl group is required for gelation, because methylating the
HA carboxyl group with diazomethane completely abolished the gelation.
The HA-Cu^{+2} gels, however, are unsuitable for ophthalmological applications
because of the toxicity of the copper ion.[108]

   Johns et al.[109] have reported that the presence of trivalent ferric ion may
significantly increase the viscosity of HA. For example, the viscosity of 1%
HA with 90% crosslinking by ferric ions reached 32,600 cps, whereas 1% HA
without ferric ions had a viscosity of 1000 cps. The maximum theoretical
binding, or 100% crosslinking, is based on one ferric ion for three repeating
disaccharide units, each carrying one carboxyl group. In addition to HA

concentration, the viscosity of the ferric ion crosslinked HA depends on the molar ratio of ferric ion present. The contribution of ferric ion to the increase of HA viscosity is significant. For instance, 5% crosslinked HA at 1.2% HA concentration has a viscosity of 1970 cps, whereas a 25% crosslinked HA at 0.9% HA concentration has a viscosity of 7660 cps. All viscosity measures were taken at the same shear rate of 1 second^{-1}. The 0.5% ferric hyaluronate was the recipe for the amber-colored Intergel indicated for use in postsurgical adhesion prevention.[110]

2.9.7 Ugi and Passerini Reactions

The use of the Ugi reaction to synthesize fluoresceinated HA was mentioned in Section 2.5.2. The Ugi reaction involves four reacting components, carboxylate, aldehyde, isonitrile, and amine. When bifunctional amino compounds such as 1,5-diaminopentane or lysine ethyl ester was used in conjunction with HA, formaldehyde and cyclohexylisocyanide, the crosslinking of HA occurred and a stable amide bond was formed. The Passerini reaction involves three types of reagents, carboxyl, aldehyde, and isonitrile. Crosslinking of HA occurred when the bifunctional aldehyde — glutaraldehyde — was used in conjunction with isonitrile. An ester bond is formed during the chemical reaction to form the hydrogel (Fig. 2.34).[111]

Ugi and Passerini reactions are simple and versatile methods for the synthesis of hydrogels based on carboxylated polysaccharides. The work by de Nooy et al.[112] compared the Ugi and Passerini reactions of a series of biopolymers containing carboxylate functions, such as HA and alginic acid and CMC. The swelling (equilibrium weight of gel in dialyzed water divided by its initial weight) of crosslinked HA with 1,5 diaminopentane was as high as 21 as compared to those of alginic acid and CMC, and so on (<10). The range of crosslinking investigated was between 5 and 10%, based on the

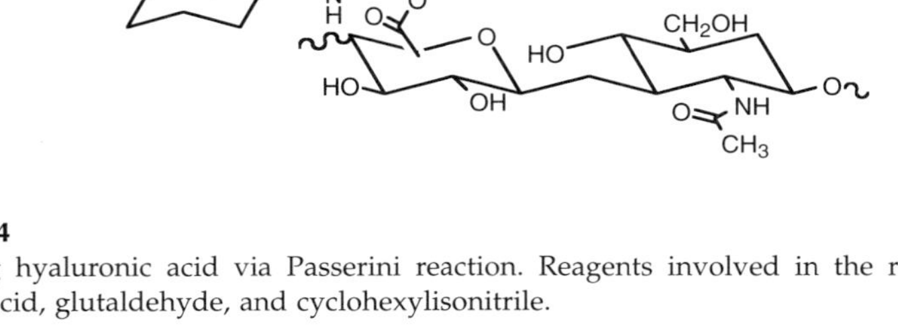

FIGURE 2.34
Crosslinking hyaluronic acid via Passerini reaction. Reagents involved in the reaction are hyaluronic acid, glutaldehyde, and cyclohexylisonitrile.

theoretical crosslinks density. The destiny was determined by the molar amount of bifunctional crosslinkers such as 1,5-diaminopentane with respect to the molar amount of carbohydrate monomers.

Interestingly, it was found that the volume of the gels prepared by Passerini reaction increased slowly after the gels were prepared, presumably because of the hydrolysis of the ester bonds formed during the crosslinking reaction. The gels from the Passerini reactions were transparent. On the contrary, the hydrogels from the Ugi reactions were stable for several months within a pH range from 1.3 to 11, at temperatures up to 90°C, as a result of the better stability of the amide bonds. Their transparency depended on the polysaccharide, the crosslinker, and the degree of crosslinking. HA and CMC tend to yield less soluble and transparent gels than alginic acid. The cyclohexyl isocyanide afforded less transparent gels than n-butyl isocyanide. The third factor, of course, was the degree of crosslinking. For example, a 5% crosslinked CMC gel was transparent, whereas an 8% crosslinked CMC was a white gel.

The quantitative analysis of the degree of crosslinking studies by de Nooy et al. included the use of ^{13}C and ^{15}N CP-MAS (crosspolarization and magic angle spinning) NMR spectroscopy. Solid-state NMR has become a valuable analytical tool for the characterization of insoluble gel or high-MW biomaterials. Pouyani et al. first used CP-MAS for characterizing the chemically crosslinked HA (by carbodiimide) at levels as low as 1% of total monomer units.[113] In the CP-MAS studies by de Nooy et al., the single pulse excitation technique was incorporated to avoid signal saturation and to improve the capacity for quantitative NMR analysis. The actual degree of crosslinking turned out to be approximately 80% of the theoretical value.[112]

2.9.8 *In Situ* Crosslinking

Reactive linear polymers may react *in situ* to form a three-dimensional network. This is particularly useful for the preparation of tissue scaffolding materials. The reactive biomaterial before *in situ* crosslinking can be prepared in injectable or moldable forms conforming to the dimension and shapes of the anatomical site where the material is applied. The material can be introduced by laproscopic devices and is therefore suitable for minimum invasive procedures. In addition, the biomaterial crosslinked at the site can produce gels with better adherence to the tissue surface.[114]

The chemistry of *in situ* crosslinking of HA needs to meet two challenges. The first is to identify the functional groups that can be crosslinked efficiently under clinical or *in vitro* conditions. For example, α, β-unsaturated esters or amides, such as acrylic esters, can be polymerized when they are exposed to light[115]; thiolated HA derivatives can self crosslink to form a disulfide bond when exposed to the air[116]; thiolated HA can conjugate with acrylic esters[114]; and HA aldehyde derivatives and HA hydrazide derivatives can crosslink by simply mixing the two.[117] The second challenge is to work out a chemical method that allows covalent bonding of such crosslinkable functions to HA.

The type of chemical methods actually used to immobilize such reactive groups to HA have been described previously in this chapter, such as conjugation reactions via hydrazides, amines and aldehydes, and so forth.

Park et al. synthesized methacrylated HA via the reaction of HA with N-3-aminopropyl methacrylamide in the presence of EDC. Then HA hydrogel was made via typical photo polymerization. HA was exposed to a xenon arc lamp at 480–520 nm, triethanolamine was used as an initiator, and eosin Y was used as a visible light sensitizer.[115] Shu et al. reported the preparation of thiol derivatives of HA (HA-DTPH) by the reaction of HA with 3,3-dithiobis(propionic hydrazide), followed by the reduction with dithiothreitol. The thiol group thus formed then readily reacted with the α, β unsaturated esters or amides, such as PEG-diacrylate or PEG-diacrylamide and so on. Unlike the polymerization reaction of acrylic derivatives, the conjugation of thiol group and acrylic esters or amides is not exothermic and is highly selective for thiol groups compared to other nucleophiles such as amine, carboxylate, or hydroxyl groups.[114,116,118,119]

2.10 Preparation of Oligosaccharide

The MW of HA oligosaccharides discussed in this section will be in the range of up to 10,000 Da, or with the number of monosaccharide units up to 50. The technology of making distinct fractions of HA oligosaccharides and the technique for a complete separation of individual fractions within that range have been developed and is available. For HA with MW in the range of 10^4 to 10^5, the HA samples are likely to be polydispersed.

2.10.1 Enzymatic Degradation

Both testicular and bacterial hyaluronidases are endohexosaminidases that hydrolyze only the N-acetylglucosaminidic bonds of HA from all sources alike. Whereas testicular hyaluronidase produces tetra-, hexa-, and octasaccharides,[120] the end product of bacterial hyaluronidase is a disaccharide.[121,122,123] In contrast to the hexosaminidases, the hyaluronidase of the leech hydrolyzes the internal glucuronidic bonds of HA and is, therefore, a β-endoglucoronidase.[124]

Nebinger et al. described a systematic way of making HA oligosaccharides with distinct units and numbers of monosaccharides.[125] As mentioned above, even-numbered oligosaccharides with a terminal nonreducing glucuronic acid were prepared from HA by degradation with bovine testicular hyaluronidase.[120] By removing the nonreducing glucuronic acid moieties from the even-numbered HA oligosaccharides with pure β-glucuronidase, odd-numbered HA oligosaccharides (hepta-, penta-, and trisaccharides) with terminal nonreducing N-acetylglucosamine were obtained. The reaction products were then separated by chromatography on DEAE-Sephacel.

Even-numbered oligosaccharides with N-acetylglucosamine at the nonreducing terminus were the products of digest by leech hyaluronidase. The treatment of such HA oligosaccharides (tetra-, hexa-, and octasaccharide) with β-N-acetylglucosaminidase A from bovine spleen afforded corresponding odd-numbered oligosaccharides with a terminal nonreducing glucuronic acid. The degradation products were isolated and separated on a preparative scale by chromatography on Dowex 1-X8 (formate form). The limitation of ion exchange chromatography in the separation of even-numbered HA oligomers (because of their identical charge and size ratio) was partially overcome by the use of organic solvent acetonitrile as eluent, which prolonged the retention time of the oligomers of HA as they passed through the amino-modified silica gel column (Fig. 2.35).

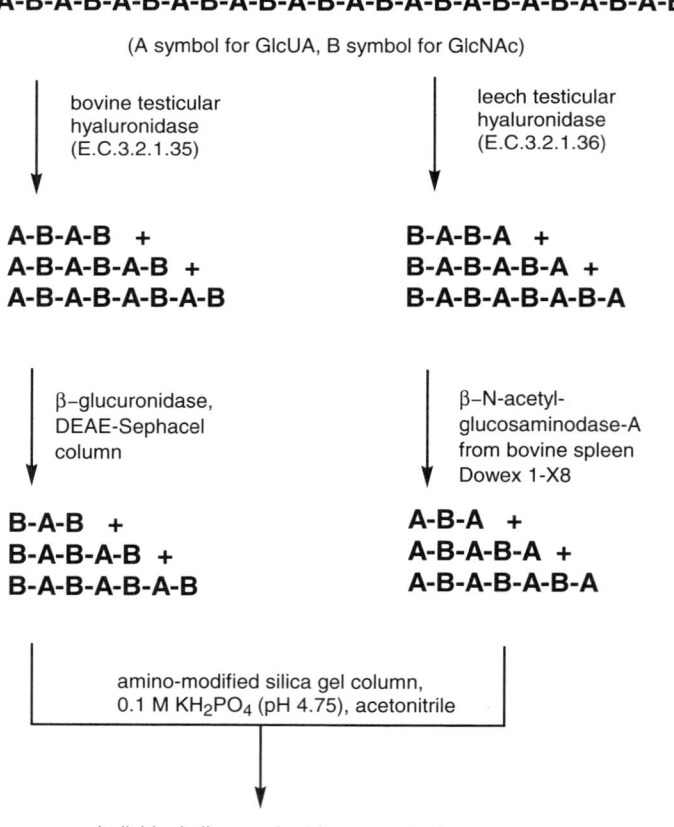

--A-B-A-B-A-B-A-B-A-B-A-B-A-B-A-B-A-B-A-B-A-B-A-B-

(A symbol for GlcUA, B symbol for GlcNAc)

bovine testicular hyaluronidase (E.C.3.2.1.35)

leech testicular hyaluronidase (E.C.3.2.1.36)

A-B-A-B +
A-B-A-B-A-B +
A-B-A-B-A-B-A-B

B-A-B-A +
B-A-B-A-B-A +
B-A-B-A-B-A-B-A

β–glucuronidase, DEAE-Sephacel column

β–N-acetyl-glucosaminodase-A from bovine spleen Dowex 1-X8

B-A-B +
B-A-B-A-B +
B-A-B-A-B-A-B

A-B-A +
A-B-A-B-A +
A-B-A-B-A-B-A

amino-modified silica gel column, 0.1 M KH_2PO_4 (pH 4.75), acetonitrile

individual oligosaccharides separated

FIGURE 2.35
Distinct oligosaccharides of hyaluronic acid can be generated by the action of selected, specific enzymes. Individual oligosaccharides, up to octasaccharide, can be separated by various methods of high-performance liquid chromatography.

Perhaps the most commonly used and widely reported methods to make HA oligomers (from HA polymer) for biological researches are those using testicular hyaluronidase. The separation methods include size-exclusion chromatography, ion-exchange chromatography, and tangential flow ultra filtration (TFUF). Examples of preparations of HA oligomers of different sizes (including separation techniques) by different researchers are 4-25 disaccharides (Sephracryl S300) by West et al.[126]; 3-16 disaccharides (G50 Sephadax) by Slevin et al.[127]; 2-15 disaccharides (G50 Sephadax) by Lokeshwar et al.[128]; 4-6 disaccharides (Biogel P-10) by Fieber et al.[129]; 3-12 disaccharides (TFUF) by Zeng et al.[130]; 4-mers to 34-mers, with homogeneous fractions from 4- to 16-mers (Bio-Gel P-6 and anion exchange), by Mahoney et al.[131]; and 2-26 disaccharides, or 4-mers to 52-mers (Dowex 1x2) by Tawada et al.[132]

Tawada's method is capable of separating "a ladder of HA oligosaccharides that vary by one repeating disaccharide" from 4-mers to 52-mers with more than 93% purity of each. A comprehensive characterization of the individual HA oligomers was accomplished by HPLC, fluorophore-assisted carbohydrate electrophoresis (FACE), electrospray-ionization mass spectrometry (ESI-MS), and NMR. From a batch of 200 g of HA, up to a gram scale of preparations of size-uniformed HA oligomers could be obtained, that were "free of protein, DNA, and endotoxin contaminants."

2.10.2   Chemical Synthesis

The success of the medical application of HA in general, and the findings that the biological behavior of HA may vary significantly according to its size in particular, have raised interest in the study of chemical synthesis of HA. In addition to enzymatic degradation and the chemoenzymatic approach, synthetic methods that afford HA oligosaccharides from disaccharide to octasaccharide may gain usefulness as their methodology continues to improve.

Carter et al. synthesized 1,4-linked HA disaccharide with a methylated hydroxyl group at the reducing end.[133] Slaghek et al. synthesized methoxyphenyl glycosides of HA di- and tetra-saccharides having either a D-glucosamine[134] or D-glucuronic acid[135] residue at the reducing end. In all those syntheses, the HA oligosaccharide backbones were first formed chemically, and the D-glucuronic acid residues were obtained by selective oxidation at C-6 of the corresponding glucose. Blatter and Jacquinet,[136] in contrast, used direct coupling of D-glucuronic acid derivatives with 2-deoxy-2-trichloroacetamido-D-glucopyranose derivatives to prepare stereo-controlled HA tetra-, hexa-, and octa-saccharides with a methyl D-glucopyranosiduronic acid residue at the reducing end.

Adamski-Werner et al. recently reported the success of gram-scale syntheses of both (1→3)-linked (7% overall yield) and (1→4)-linked HA disaccharides (12% overall yield).[137] Both syntheses installed the uronic functionality by TEMPO oxidation (TEMPO is 2,2,6,6-tetramethyl-1-piperidinyloxy free radical [2564-83-2]).

Unlike the synthesis of peptides or oligonucleotides, the synthesis of complex carbohydrate is laborious as a result of the multifunctional groups of each building block and is normally performed in highly specialized laboratories. The traditional solution-phase chemistry remains time consuming despite the continuing effort to improve it. A new area of active research and development for the synthesis of HA oligosaccharides is the automated solid phase synthesis.[138]

Palmacci et al. developed the first automated oligosaccharide synthesizer that works in a fashion analogous to the peptide synthesizer.[139] The growing oligosaccharide chain is attached to the polystyrene resin through an octene-diol linker that can be cleaved from the resin to afford a pentenyl glycoside. The synthesis requires only the suitably protected mono- or di-saccharide glycosyl donors, and the synthesizer can accomplish glycosylation and deprotection reactions effectively. The method gave higher yield and worked about 20 times faster than the solution-based methods for a number of important carbohydrates, including a proteoglycan linkage-region tetrasaccharide. Palmacci and Seeberger, in pursuing the program of automated synthesis of HA, have succeeded in the preparation of two differentially protected GluNAc-β(1→4)-GlcUA and GlcUA-β(1→3)-GluNac disaccharide modules.[140] The glucuronic acid moiety was transformed from glucose by periodic/chromium trioxide, a mild oxidation method that yielded as high as 70% of the glucuronic acid.

2.10.3   Chemoenzymatic Synthesis

HA can be prepared by HA synthetase and relatively inexpensive uridine diphospho (UDP)–sugar substrates, UDP-GlcNAc and UDP-GlcUA. But the process led to high-MW HA (5×10^5 Da).[141] Recombinant technology has been developed to synthesize HA oligosaccharide with limited and precise lengths.[142,143]

The enzyme HA synthase (HAS) belongs to the class of glycosyltransferases (or transferases) that are responsible for the synthesis of the alternating and repeating saccharide chain of HA. Transferases use UDP-sugar precursors and metal cofactors (Mg or its ion) to synthesize the polysaccharide (Eq. 2.1). The reaction usually proceeds at near neutral pH ($n = 25$ to 10,000):

$$n \text{ UDP-GlcUA} + n \text{ UDP-HexNAc} \rightarrow 2n \text{ UDP} + [\text{GlcUA-GlcNAc}]_n \quad (2.1)$$

Most of the known glycosyltransferases transfer only one type of monosaccharide to an acceptor molecule. The native forms of HA synthases, however, are dual-action glycosyltransferases that add rapidly two different sugars, GlcUA and GlcNAc, to form a polymer chain and are therefore unsuitable for production of short, defined oligosaccharides. It is not practical to rely on kinetic control of the process such as reducing substrate concentration, temperature, and reaction time to control the size of the HA oligosaccharide precisely and reproducibly.

Pasteurella multocida HA synthase (pmHAS) can be converted by mutation of the aspartate residue into two single-action glycosyltransferases (glucuronic acid transferase and *N*-acetylglucosamine transferase). A lesion in one active site leaves the activity of the other unmutated active site relatively unaffected. Therefore, separate reactions, as represented in Eqs. (2.2) and (2.3) are now possible.

$$\text{UDP-GlcNAc} + [\text{GlcUA-GlcNAc}]n \rightarrow \text{UDP} + \text{GlcNAc-}[\text{GlcUA-GlcNAc}]n$$

(2.2)

$$\text{UDP-GlcUA} + [\text{GlcNAc-GlcUA}]n \rightarrow \text{UDP} + \text{GlcUA-}[\text{GlcNAc-GlcUA}]n$$

(2.3)

When the two resulting transferases, a β3GlcNAc-transferase and a β4GlcUA-transferase, are purified and immobilized individually onto solid supports, the enzyme reactors used in an alternating fashion can produce an extremely pure, single-length HA oligosaccharide up to HA20, without purification of the intermediates.

The reason that pmHAS was selected for the synthesis of HA oligosaccharides with precise lengths has to do with its unique structural characteristics. There are two classes of HA synthases. Class I includes the Gram-positive streptococcal, vertebrate, and viral HAS, whereas the class II enzyme has its only member — Gram-negative *Pasteurella*.[144] The *Pasteurella* enzyme contains a carboxy-terminal region that is membrane bound, and the deletion of the region results in a soluble, cytoplasmic form of the enzyme.[145] Only the recombinant pmHAS, not streptococcal or vertebrate enzymes, has shown the ability to perform reactions to elongate HA-oligosaccharide acceptors exogenously supplied.

2.11 Sulfation of HA

Various methods of sulfation of HA have been reported. Balazs et al.[146] prepared polysulfated HA by suspending dry HA in a mixture of chlorosulfonic acid and pyridine. Chang et al.[147] adopted that method and prepared sulfated HA with a sulfate/hexosamine molar ratio of 3.9. In the control reaction, Chang used triethylamine and sulfur trioxide according to the procedures for sulfation of starch described by Whistler and Spencer,[148] which resulted in lower degrees of sulfation of 0.4, 0.6, and 1.2. The polysulfated HA demonstrated certain inhibition of TNF-α and TNF-β, whereas the

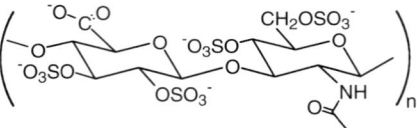

FIGURE 2.36
Polysulfated hyaluronic acid.

naturally occurring sulfated polysaccharides such as chondroitin sulfate was not effective.

Barbucci et al. carried out sulfation of HA with SO_3-pyridine in DMF, and HA was first converted to its TBA salt before sulfation. The degree of sulfation depended on the ratio of SO_3/HA disaccharide units, ranging from 1 to 4. The 6-OH group of HA was the first to be sulfated, and when the ratio increased to 4, all hydroxyl groups were sulfated. The conclusion on the structure of the sulfated HA was substantiated by elemental analysis of C, H, N, and S and by NMR spectroscopy. For instance, the 6-OH sulfation was evidenced by the "low field shift" (from 61.1 to 67.7 ppm) of the glucosamine C_6 signal. The biological effect of sulfated HA was evaluated, including anticoagulant and antiinflammatory activities (Fig. 2.36).

2.12 Summary of Methods of Chemical Modification

Chemical modifications of HA via its functional groups were developed with certain practical purposes in mind: to chemically label HA with a fluorescent or radioactive group as a tracer in many *in vivo* studies of HA; to make new HA derivatives, including crosslinked HA, as novel biomaterials with altered chemical, physical, and biological properties; and to synthesize HA–drug conjugates as prodrug or to modify certain pharmacological properties of the targeted drugs.

There are three functional groups of HA that are often the targets for chemical modification: carboxyl, hydroxyls, and aldehyde groups. The N-acetyl group on GlcNAc needs to be deacetylated to generate a reactive amino group as the reaction site for further modification of HA. The deacetylation step is likely to degrade the HA chains, and therefore its use is quite limited. The glycosidic bond is the chemical bond one wants to preserve, unless the purpose includes the reduction of the MW of HA.

Several methods have proven to be useful in the chemical modification of HA. Table 2.2 summarizes those methods mentioned previously in this chapter. The table is organized according to the HA functional groups targeted, the reagents used, and the chemical linkage formed as the result of the chemical reactions.

TABLE 2.2

Summary of Chemical Modification via Functional Groups

Targeted Hyaluronic Acid Functions	Reagent(s)	Formed Hyaluronic Acid Linkage(s)
Carboxyl	Carbodiimide	Acylurea
Predominantly in its ionized	Hydrazine/carbodiimide	Hydrazide
form under neutral, alkaline,	Amine/isonitrile/aldehyde (Ugi)	Amide
or slightly acidic conditions	Amine/NHS/carbodiimide	Amide
	Isonitrile/aldehyde (Passerini)	Ester
	HA hydroxyl in self x-linking	Ester
Carboxyl/TBA Complex	Alkyl (carbonyl) halides	Ester
	NHS-diphenyl phosphate	(Active) Ester
Hydroxyls	Divinylsulfone	Ether
GlcNAc (6 and 4-OH)	Amine (-OH activated by	Carbamate
GlcUA (2 and 3-OH)	cyanogen bromide)	
	Epoxide, OH⁻	Ether
	Epoxide, H⁺	Ester
	Isothiocyanate	Thiocarbamate
	Sulfating agent	Sulfate
Aldehyde	Amine/NaBH₄,	Amine
At reducing end, or at vicinal	(reductive amination)	
2 and 3 position of GlcUA by	Hydrazine	Hydrazone
peroxide reaction		

References

1. Weissman, B. and Meyer, K., The structure of hyalobiuronic acid and of hyaluronic acid from umbilical cord, *J. Am. Chem. Soc.*, 76, 1753–1757 (1954).
2. Flowers, H. M. and Jeanloz, R. W., The synthesis of the repeating units of hyaluronic acid, *Biochemistry*, 3, 123–125 (1964).
3. Jeanloz, R. W. and Jeanloz, D. A., The degradation of hyaluronic acid by methanolysis, *Biochemistry*, 3, 121–122 (1964).
4. Murphy, D., Pennock, C. A., and London, K. J., Gas-liquid chromatographic measurement of glucosamine and galactosamine content of urinary glycosaminoglycans, *Clin. Chim. Acta*, 53, 145–152 (1974).
5. Radhakrishnamurthy, B., Dalferes, E. R. Jr., and Berenson, G. S., Determination of hexosamines by gas-liquid chromatography, *Anal. Biochem.*, 17, 545–550 (1966).
6. Montgomery, R. and Nag, S., Periodate oxidation of hyaluronic acid, *Biochim. Biophys. Acta*, 74, 300–302 (1963).
7. Scott, J. E., Periodate oxidation, pKa, and conformation of hexuronic acids in polyuronides and mucopolysaccharides, *Biochim. Biophys. Acta*, 170, 471–473 (1968).
8. Scott, J. E. and Tigwell, M. J., The influence of the intrapolymer environment on periodate oxidation of uronic acids in polyuronides and glycosaminoglycuronans, *Biochem. Soc. Trans.*, 3, 662–664 (1975).
9. Kennedy, J. F. and White, C. A., *Bioactive Carbohydrates in Chemistry, Biochemistry, and Biology.* Ellis Horwood, New York, 1983.

10. Dische, Z., A new specific color reaction of hexuronic acid, *J. Biol. Chem.*, 167, 189–198 (1946).
11. Elson, L. A. and Morgan, W. T. J., A colorimetric method for the determination of glucosamine and chondrosamine, *Biochem. J.*, 27, 1824–1828 (1933).
12. Blix, G., The determination of hexosamines according to Elson and Morgan, *Acta Chem. Scand.*, 2, 467–473 (1948).
13. Morgan, W. T. J. and Elson, L. A., A colorimetric method for the determination of N-acetylglucosamine and N-acetylchondrosamine, *Biochem. J.*, 28, 988–995 (1934).
14. Aminoff, D., Morgan, W. T. J., and Watkins, W. M., The action of dilute alkali on the N-acetylhexosamines and the specific blood-group mucoids, *Biochem. J.*, 51, 379–389 (1952).
15. White, T., Studies in the amino-sugars. Part II. The action of dilute alkali solution on acylglucosamine, *J. Chem. Soc.*, 428–437 (1940).
16. Reissig, J. L., Strominger, J. L., and Leloir, L. F., A modified colorimetric method for the estimation of N-acetylamino sugars, *J. Biol. Chem.*, 217, 959–966 (1955).
17. Kuhn, R. and Krüger, R., Das Chromogen III der Morgan-Elson Reaktion, *Chem. Ber.*, 90, 264–277 (1957).
18. Beau, J.-M., Rollin, P., and Sinay, P., Structure du chromogène I de la rèaction de Morgen Elson, *Carbohydr. Res.*, 53, 187–195 (1977).
19. Benson, R. L., Testing proposed reaction mechanisms with compounds bound to solid supports, *J. Org. Chem.*, 40, 1647–1649 (1975).
20. Muckenschnabel, I., Bernhardt, G., Spruss, T., Dietl, B., and Buschauer, A., Quantitation of hyaluronidases by the Morgan-Elson reaction: comparison of the enzyme activities in the plasma of tumor patients and healthy volunteers, *Cancer Lett.*, 131, 13–20 (1998).
21. Whistler, R. L. and BeMiller, J. B., Alkaline degradation of polysaccharides, *Adv. Carbohydrate Chem. Biochem.*, 13, 289–329 (1958).
22. Kennedy, J. F. and White, C. A., in *Comprehensive Organic Chemistry, The Synthesis and Reactions of Organic Compounds*, Halslam, E., Ed., Pergamon, Oxford, 1979, Vol. 5, pp. 766–768.
23. Kuo, J. W., in *Synthesis and Properties of Hyaluronic Acid Modified by Designed Carbodiimides, Department of Chemistry*, State University of New York, Stony Brook, 1989, pp. 1–131.
24. Jennings, H. J., Roy, R., and Michon, F., Determinant specificities of the groups B and C polysaccharides of Neisseria meningitidis, *J. Immunol.*, 134, 2651–2657 (1985).
25. Jennings, H. J., Roy, R., and Gamian, A., Induction of meningococcal group B polysaccharide-specific IgG antibodies in mice by using an N-propionylated B polysaccharide-tetanus toxoid conjugate vaccine, *J. Immunol.*, 137, 1708–1713 (1986).
26. Linker, A., Meyer, K., and Hoffman, P., The production of unsaturated uronides by bacterial hyaluronidases, *J. Biol. Chem.*, 219, 13–25 (1956).
27. Shively, J. E., Taylor, R. L., and Conrad, H. E., in *The Methodology of Connective Tissue Research*, Hall, D. A., Ed., Joynson-Bruvvers, Oxford, 1976.
28. Moggridge, R. C. G. and Neuberger, A., Methylglycosaminide: its structure, and the kinetics of its hydrolysis by acids, *J. Chem. Soc.*, 745–750 (1938).

29. Inoue, Y. and Nagasawa, K., Preparation, by chemical degradation of hyaluronic acid, of a series of even- and odd-numbered oligosaccharides having a 2-acetamido-2-deoxy-glucose and a glucuronic acid residue, respectively, at the reducing end, *Carbohydr. Res.*, 141, 99–110 (1985).

30. Rånby, B. G. and Marchessault, R. H., Inductive effects in the hydrolysis of cellulose chains. *J Polymer Sci.*, 36, 561–564 (1959).

31. BeMiller, J. N., Acid-catalysed hydrolysis of glycosides, *Adv. Carbohydrate Chem. Biochem.*, 22, 25–108 (1967).

32. Bochkov, A. F. and Zaikov, G. E., in *Chemistry of the O-Glycosidic Bond*, Pergamon, Oxford, 1979, pp. 177–201.

33. Roy, N. and Timell, T. E., The acid hydrolysis of glycosides: XI. Effect of pH on the hydrolysis of acid disaccharides, *Carbohydr. Res.*, 7, 17–20 (1968).

34. Pigman, W. and Rizvi, S., Hyaluronic acid and the ORD reaction, *Biochem. Biophys. Res. Commun.*, 1, 39–43 (1959).

35. Gilbert, D. L., Gerschman, R., Cohen, J., and Sherwood, W., The influence of high oxygen pressures on the viscosity of solutions of sodium desoxyribonucleic acid and of sodium alginate, *J. Am. Chem. Soc.*, 79, 5677–5680 (1957).

36. Swann, D. A., Studies on the structure of hyaluronic acid. Characterization of the product formed when hyaluronic acid is treated with ascorbic acid, *Biochem. J.*, 114, 819–825 (1969).

37. Cleland, R. L., Stoolmiller, A. C., Roden, L., and Laurent, T. C., Partial characterization of reaction products formed by the degradation of hyaluronic acid with ascorbic acid, *Biochim. Biophys. Acta*, 192, 385–394 (1969).

38. Uchiyama, H., Dobashi, Y., Ohkouchi, K., and Nagasawa, K., Chemical change involved in the oxidative reductive depolymerization of hyaluronic acid, *J. Biol. Chem.*, 265, 7753–7759 (1990).

39. Harris, M. J., Herp, A., and Pigman, W., Metal catalysis in the depolymerization of hyaluronic acid by autoxidant, *J. Am. Chem. Soc.*, 94, 7570–7572 (1972).

40. Betts, W. H. and Cleland, L. G., Effect of metal chelators and antiinflammatory drugs on the degradation of hyaluronic acid, *Arthritis Rheum.*, 25, 1469–1476 (1982).

41. Halliwell, B., Superoxide-dependent formation of hydroxyl radicals in the presence of iron chelates: is it a mechanism for hydroxyl radical production in biochemical systems? *FEBS Lett.*, 92, 321–326 (1978).

42. McCord, J. M., Free radicals and inflammation: production of synovial fluid by superoxide dismutase, *Science*, 185, 529–531 (1974).

43. Wedlock, D. J., Phillips, G. O., Davies, A., Gormally, J., and Wyn-Jones, E., Depolymerization of sodium hyaluronate during freeze drying, *Int. J. Biol. Macromol.*, 5, 186–188 (1983).

44. Tokita, Y., Oshima, K., and Okamoto, A., Degradation of hyaluronic acid during freeze drying, *Polymer Degradation Stability*, 55, 159–164 (1997).

45. Balazs, E. A. and Laurent, T. C., The viscosity function of hyaluronic acid as a polyelectrolyte, *J. Polymer. Sci.*, 6, 665–668 (1951).

46. Balazs, E. A., Laurent, T. C., Howe, A. F., and Varga, L., Irradiation of mucopolysaccharides with ultraviolet light and electrons, *Radiat. Res.*, 11, 149–164 (1959).

47. Balazs, E. A., Davies, J. V., Phillips, G. O., and Young, M. D., Transient intermediates in the radiolysis of hyaluronic acid, *Radiat. Res.*, 31, 243–255 (1967).

48. Caputo, A., Depolymerization of hyaluronic acid by x-rays, *Nature*, 179, 1133–1134 (1957).

49. Lowry, K. M. and Beavers, E. M., Thermal stability of sodium hyaluronate in aqueous solution, *J. Biomed. Mater. Res.*, 28, 1239–1244 (1994).

50. Miyazaki, T., Yomota, C., and Okada, S., Ultrasonic depolymerization of hyaluronic acid, *Polymer Degradation Stability*, 74, 77–85 (2001).

51. Laurent, T. C., Dahl, I. M., Dahl, L. B., Engstrom-Laurent, A., Eriksson, S., Fraser, J. R., Granath, K. A., Laurent, C., Laurent, U. B., Lilja, K., The catabolic fate of hyaluronic acid, *Connect. Tissue Res.*, 15, 33–41 (1986).

52. Fraser, J. R., Laurent, T. C., Engstrom-Laurent, A., and Laurent, U. G., Elimination of hyaluronic acid from the blood stream in the human, *Clin. Exp. Pharmacol. Physiol.*, 11, 17–25 (1984).

53. Underhill, C. B. and Toole, B. P., Binding of hyaluronate to the surface of cultured cells, *J. Cell Biol.*, 82, 475–484 (1979).

54. Baxter, E., Fraser, J. R., and Harris, G. S., Fractionation and recovery of secretions of synovial cells synthesized in culture with radioactive precursors, *Ann. Rheum. Dis.*, 32, 35–40 (1973).

55. Wilzbach, K. E., Tritium-labeling by exposure of organic compounds to tritium gas, *J. Am. Chem. Soc.*, 79, 1013 (1957).

56. Highsmith, S. and Chipman, D. M., Preparation of tritium-labeled hyaluronic acid oligomers and their use in enzyme studies, *Anal. Biochem.*, 61, 557–566 (1974).

57. Hook, M., Riesenfeld, J., and Lindahl, U., N-[3H]acetyl-labeling, a convenient method for radiolabeling of glycosaminoglycans, *Anal. Biochem.*, 119, 236–245 (1982).

58. Swann, D. A., Silver, F. H., Sotman, S. L., and Hermann, H., Measurements of reducing end groups on bovine vitreous-humour hyaluronic acid by reaction with ^{14}C cyanide, *Biochem. J.*, 207, 409–414 (1982).

59. Orlando, P., Binaglia, L., De Feo, A., Orlando, M., Trenta, R., and Trevisi, R., An improved method for hyaluronic acid radioiodination, *J. Labelled Compounds Radiopharmaceut.*, 36, 855–859 (1995).

60. Gustafson, S., Bjorkman, T., and Westlin, J. E., Labelling of high molecular weight hyaluronan with 125I-tyrosine: studies *in vitro* and *in vivo* in the rat, *Glycoconj. J.* 11, 608–613 (1994).

61. Westerberg, G., Bergstrom, M., Gustafson, S., Lindqvist, U., Sundin, A., and Langstrom, B., Labelling of polysaccharides using [11C]cyanogen bromide. *In vivo* and *in vitro* evaluation of 11C-hyaluronan uptake kinetics, *Nucl. Med. Biol.*, 22, 251–256 (1995).

62. Orlando, M., De Sanctis, G., Giodano, C., Valle, G., La Bua, R., and Ghidoni, R., Tritium labeled hyaluronic acid derivative, *J. Labelled Compounds Radiopharmaceut.*, 22, 961–969 (1985).

63. Dmitriev, B. A., Knirel Y. A., and Kochetkov, N. K., Selective cleavage of glycosidic linkages: studies with the model compound benzyl 2-acetamido-2-deoxy-3-O-β-D-galactopyranosyl-α-D-glucopyranoside, *Carbohydr. Res.*, 29, 451–457 (1973).

64. Robert, L., Mednieks, M., and Winzler, R. J., Reaction of [^{14}C]cyanide with hexosamines, acid polysaccharides, and some related compounds, *Biochim. Biophys. Acta*, 63, 327–333 (1962).

65. Raja, R. H., LeBoeuf, R. D., Stone, G. W., and Weigel, P. H., Preparation of alkylamine and 125I-radiolabeled derivatives of hyaluronic acid uniquely modified at the reducing end, *Anal. Biochem.*, 139, 168–177 (1984).

66. Wong, S. S., *Chemistry of Protein Conjugation and Crosslinking*, CRC Press, Boca Raton, 1993.

67. Gustafson, S., Forsberg, N., McCourt, P., Wikstrom, T., Bjorkman, T., and Lilja, K., in *First International Workshop on Hyaluronan in Drug Delivery*, Willoughby, D. A., Ed., Royal Society of Medicine Press, Toronto, 1992, pp. 43–59.
68. de Belder, A. N. and Wik, K. O., Preparation and properties of fluorescein-labelled hyaluronate, *Carbohydr. Res.*, 44, 251–257 (1975).
69. Ugi, J., *Isonitrile Chemistry*, Academic Press, London, 1971.
70. Ogamo, A., Matsuzaki, K., Uchiyama, H., and Nagasawa, K., Preparation and properties of fluorescent glycosaminoglycuronans labeled with 5-aminofluorescein, *Carbohydr. Res.*, 105, 69–85 (1982).
71. Luo, Y. and Prestwich, G. D., Hyaluronic acid-N-hydroxysuccinimide: a useful intermediate for bioconjugation, *Bioconjug. Chem.*, 12, 1085–1088 (2001).
72. Pouyani, T. and Prestwich, G. D., Biotinylated hyaluronic acid: a new tool for probing hyaluronate-receptor interactions, *Bioconjug. Chem.*, 5, 370–372 (1994).
73. Della Valle, F. and Romeo, A., Esters of hyaluronic acid, U.S. Patent 4,851,521, (1989).
74. Yu, Q. and Toole, B. P., Biotinylated hyaluronan as a probe for detection of binding proteins in cells and tissues, *Biotechniques*, 19, 122–124, 126–129 (1995).
75. Yang, B., Yang, B. L., and Goetinck, P. F., Biotinylated hyaluronic acid as a probe for identifying hyaluronic acid-binding proteins, *Anal. Biochem.*, 228, 299–306 (1995).
76. Frost, G. I. and Stern, R., A microtiter-based assay for hyaluronidase activity not requiring specialized reagents, *Anal. Biochem.*, 251, 263–269 (1997).
77. Bragg, P. D. and Hou, C., Subunit composition, function, and spatial arrangement in the Ca^{2+}- and Mg^{2+}-activated adenosine triphosphatases of *Escherichia coli* and *Salmonella typhimurium*, *Arch. Biochem. Biophys.*, 167, 311–321 (1975).
78. Lomant, A. J. and Fairbanks, G., Chemical probes of extended biological structures: synthesis and properties of the cleavable protein crosslinking reagent [35S]dithiobis(succinimidyl propionate), *J. Mol. Biol.*, 104, 243–261 (1976).
79. Hermanson, G. T., *Bioconjugate Techniques*, Academic Press, San Diego, CA, 1996, pp. 139–141.
80. Kuo, J. W., Swann, D. A., and Prestwich, G. D., Chemical modification of hyaluronic acid by carbodiimides, *Bioconjug. Chem.*, 2, 232–241 (1991).
81. Hamilton, R., Fox, E. M., Acharya, R. A., and Walts, A. E., Water insoluble derivatives of hyaluronic acid, U.S. Patent 4,937,270 (1990).
82. Ruiz-Perez, B., Cisneros, R. L., Matsumoto, T., Miller, R. J., Vasios, G., Calias, P., and Onderdonk, A. B., Protection against lethal intra-abdominal sepsis by 1- (3-dimethylaminopropyl)-3-ethylurea, *J. Infect. Dis.*, 188, 378–387 (2003).
83. Ruiz-Perez, B., Cisneros, R. L., Matsumoto, T., Miller, R. J., Vasios, G., Calias, P., Onderdonk, A. B., Tzianabos, A. O., Gershkovich, J., Johnson, J., Burns, J. W., and Stavesk, M., Therappeutic utility of chemically modified hyaluronan, Poster 62 Conference, Hyaluronan 2003, Cleveland, OH, 2003.
84. Vercruysse, K. P. and Prestwich, G. D., Hyaluronate derivatives in drug delivery, *Crit. Rev. Ther. Drug Carrier Syst.*, 15, 513–555 (1998).
85. Pouyani, T. and Prestwich, G. D., Functionalized derivatives of hyaluronic acid oligosaccharides: drug carriers and novel biomaterials, *Bioconjug. Chem.*, 5, 339–347 (1994).
86. Luo, Y. and Prestwich, G. D., Synthesis and selective cytotoxicity of a hyaluronic acid-antitumor bioconjugate, *Bioconjug. Chem.*, 10, 755–763 (1999).
87. Luo, Y., Ziebell, M. R., and Prestwich, G. D., A hyaluronic acid-taxol antitumor bioconjugate targeted to cancer cells, *Biomacromolecules*, 1, 208–218 (2000).

88. Balazs, E. A., Leshchiner, A., Leshchiner, A., and Band, P., Chemically modified hyaluronic acid preparation and method of recovery thereof from animal tissues, U.S. Patent 4,713,448 (1987).

89. Balazs, E. A. and Leshchiner, A., Crosslinked gels of hyaluronic acid and products containing such gels, U.S. Patent 4,582,865 (1986).

90. Balazs, E. A. and Denlinger, J. L., Viscosupplementation: a new concept in the treatment of osteoarthritis, *J. Rheumatol. Suppl.*, 39, 3–9 (1993).

91. Band, A. P., Hyaluronan derivatives: chemistry and clinical applications, in *The Chemistry, Biology, and Medical Applications of Hyaluronan and Its Derivatives*, Laurent, T. C., Ed., Portland Press, London, 1998, pp. 33–42.

92. Porath, J., General methods and coupling procedures, *Methods Enzymol.*, 34, 13–30 (1974).

93. Porath, J. and Axen, R., Immobilization of enzymes to agar, agarose, and Sephadex supports, *Methods Enzymol.*, 44, 19–45 (1976).

94. Vercruysse, K. P., Marecak, D. M., Marecek, J. F., and Prestwich, G. D., Synthesis and *in vitro* degradation of new polyvalent hydrazide crosslinked hydrogels of hyaluronic acid, *Bioconjug. Chem.*, 8, 686–694 (1997).

95. Smith, P. A. S., *The Curtius Reaction. Organic Reactions*, Adams, R., Bachmann, W. E., Fieser, L. F., Johnsons, J. R., and Snyder, H. R., Eds., Wiley, New York, 1946, Vol. 3, pp. 366–367.

96. Lutter, L. C., Ortanderl, F., and Fasold, H., The use of a new series of cleavable protein-crosslinkers on the *Escherichia coli* ribosome, *FEBS Lett.*, 48, 288–292 (1974).

97. Tomalia, D. A., Baker, H., Dewald, J., Hall, M., Kallos, G., Martin, S., Roeck, J., Ryder, J., and Smith, P. J., Dendritic macromolecules: synthesis of starburst dendrimers, *Macromolecules*, 19, 2466–2468 (1986).

98. Luo, Y., Kirker, K. R., and Prestwich, G. D., Crosslinked hyaluronic acid hydrogel films: new biomaterials for drug delivery, *J. Control Release*, 69, 169–184 (2000).

99. de Belder, A. N. and Malson, T., Means for preventing such adhesion, and process for producing said means, U.S. Patent 4,886,787, (1989).

100. Laurent, T., Hellsing, K., and Gelotte, B., Crosslinked gels of hyaluronic acid, *Acta Chem. Scand.*, 18, 274–275 (1964).

101. Agerup, B., Polysaccharide gel composition, U.S. Patent 5,827,937 (1998).

102. Kurzer, F. and Douraghi-Zadeh, K., Advances in the chemistry of carbodiimides, *Chem. Rev.*, 67, 107–152 (1967).

103. Mukaiyama, T., Usui, M., Shimada, E., and Saigo, K., A convenient method for the synthesis of carboxylic esters, *Chem. Lett.*, 1045–1048 (1975).

104. Saigo, K., Usui, M., Kikuchi, K., Shimada, E., and Mukaiyama, T., New method for the preparation of carboxylic esters, *Bull. Chem. Soc. Japan.*, 50, 1863–1866 (1977).

105. Della Valle, F. and Romeo, A., Crosslinked carboxy polysaccharides, U.S. Patent 5,676,964 (1997).

106. Schmut, O. and Hofmann, H., Preparation of gels from hyaluronate solutions, *Graefes Arch. Clin. Exp. Ophthalmol.*, 218, 311–314 (1982).

107. Balazs, E. A., Sediment volume and viscoelastic behavior of hyaluronic acid, *Fed. Proc.*, 25, 1817–1822 (1966).

108. Malson, T. and Lindqvist, B. L., Gel of crosslinked hyaluronic acid for use as a vitreous humor substitute, U.S. Patent 4,716,154 (1987).

109. Johns, D., Rodgers, K., Donahue, W., Kiorpes, T., and di Zerega, G. S., Reduction of adhesion formation by postoperative administration of ionically crosslinked hyaluronic acid, *Fertil. Steril.*, 68, 37–42 (1997).

110. Huang, W. J., Johns, D. B., and Kronenthal, R. L., Ionically crosslinked carboxyl-containing polysaccharides for adhesion prevention, U.S. Patent 5,532,221, Biomedical, Inc., Chaska, MN, (1996).

111. Crescenzi, V., Tomaci, M., and Francescangeli, A., New routes to hyaluronan-based networks and supramolecular assemblies, in *Redefining Hyaluronan*, Abatangelo, G. and Weigel, P. H., Eds., Elsevier, Amsterdam, 2000, pp. 173–180.

112. de Nooy, A. E., Capitani, D., Masci, G., and Crescenzi, V., Ionic polysaccharide hydrogels via the Passerini and Ugi multicomponent condensations: synthesis, behavior, and solid-state NMR characterization, *Biomacromolecules*, 1, 259–267 (2000).

113. Pouyani, T., Kuo, J.-W., Harbison, G. S., and Prestwich, G. D., Solid-state NMR of N-acylureas derived from the reaction of hyaluronic acid with isotopically-labeled carbodiimides, *J. Am. Chem. Soc.*, 114, 5972–5976 (1992).

114. Shu, X. Z., Liu, Y., Palumbo, F. S., Luo, Y., and Prestwich, G. D., *In situ* crosslinkable hyaluronan hydrogels for tissue engineering, *Biomaterials*, 25, 1339–1348 (2004).

115. Park, Y. D., Tirelli, N., and Hubbell, J. A., Photopolymerized hyaluronic acid-based hydrogels and interpenetrating networks, *Biomaterials*, 24, 893–900 (2003).

116. Shu, X. Z., Liu, Y., Palumbo, F., and Prestwich, G. D., Disulfide-crosslinked hyaluronan-gelatin hydrogel films: a covalent mimic of the extracellular matrix for *in vitro* cell growth, *Biomaterials*, 24, 3825–3834 (2003).

117. Jia, X., Colombo, G., Padera, R., Langer, R., and Kohane, D. S., Prolongation of sciatic nerve blockade by *in situ* crosslinked hyaluronic acid, *Biomaterials*, 25, 4797–4804 (2004).

118. Lutolf, M. P., Tirelli, N., Cerritelli, S., Cavalli, L., and Hubbell, J. A., Systematic modulation of Michael-type reactivity of thiols through the use of charged amino acids, *Bioconjug. Chem.*, 12, 1051–1056 (2001).

119. Friedman, M., Cavins, J. F., and Wall, J. S., Relative nucleophilic reactivities of amino groups and mercaptide ions in addition reactions with α, β-unsaturated compounds, *J. Am. Chem. Soc.*, 87, 3672–3682 (1965).

120. Rapport, M. M., Meyer, K., and Linker, A., Analysis of the products formed on hydrolysis of hyaluronic acid by testicular hyaluronidase, *J. Am. Chem. Soc.*, 73, 2416–2420 (1951).

121. Rapport, M. M., Linker, A., and Meyer, K., The hydrolysis of hyaluronic acid by pneumococcal hyaluronidase, *J. Biol. Chem.*, 192, 283–291 (1951).

122. Linker, A. and Meyer, K., Production of unsaturated uronides by bacterial hyaluronidases, *Nature*, 174, 1192–1194 (1954).

123. Cabezas, J. A., Reglero, A., De Pedro, A., Diez, T., and Calvo, P., Hydrolysis of natural and synthetic substrates by alpha-L-fucosidase, beta-D-glucuronidase, and beta-N-acetylhexosaminidase purified from molluscs, *Int. J. Biochem.*, 13, 389–393 (1981).

124. Linker, A., Meyer, K., and Hoffman, P., The production of hyaluronate oligosaccharides by leech hyaluronidase and alkali, *J. Biol. Chem.*, 235, 924–927 (1960).

125. Nebinger, P., Koel, M., Franz, A., and Werries, E., High-performance liquid chromatographic analysis of even- and odd-numbered hyaluronate oligosaccharides, *J. Chromatography*, 265, 19–25 (1983).

126. West, D. C., Hampson, I. N., Arnold, F., and Kumar, S., Angiogenesis induced by degradation products of hyaluronic acid, *Science*, 228, 1324–1326 (1985).

127. Slevin, M., Kumar, S., and Gaffney, J., Angiogenic oligosaccharides of hyaluronan induce multiple signaling pathways affecting vascular endothelial cell mitogenic and wound healing responses, *J. Biol. Chem.*, 277, 41046–41059 (2002).

128. Lokeshwar, V. B. and Selzer, M. G., Differences in hyaluronic acid-mediated functions and signaling in arterial, microvessel, and vein-derived human endothelial cells, *J. Biol. Chem.*, 275, 27641–27649 (2000).

129. Fieber, C., Baumann, P., Vallon, R., Termeer, C., Simon, J. C., Hofmann, M., Angel, P., Herrlich, P., and Sleeman, J. P., Hyaluronan-oligosaccharide-induced transcription of metalloproteases, *J. Cell Sci.*, 117, 359–367 (2004).

130. Zeng, C., Toole, B. P., Kinney, S. D., Kuo, J. W., and Stamenkovic, I., Inhibition of tumor growth *in vivo* by hyaluronan oligomers, *Int. J. Cancer*, 77, 396–401 (1998).

131. Mahoney, D. L., Aplin, R. T., Calabro, A., Hascall, V. C., and Day, A. J., Novel methods for the preparation and characterization of hyaluronan oligosaccharides of defined length, *Glycobiology*, 11, 1025–1033 (2001).

132. Tawada, A., Masa, T., Oonuki, Y., Watanabe, A., Matsuzaki, Y., and Asari, A., Large-scale preparation, purification, and characterization of hyaluronan oligosaccharides from 4-mers to 52-mers, *Glycobiology*, 12, 421–426 (2002).

133. Carter, M. B., Petillo, P. A., Anderson, L., and Lerner, L. E., The 1,4-linked disaccharide of hyaluronan: synthesis of methyl 2-acetamido-2-deoxy-beta-D-glucopyranosyl-(14)-beta-D-glucopyrano sid uronic acid, *Carbohydr. Res.*, 258, 299–306 (1994).

134. Slaghek, T. M., Nakahara, Y., Ogawa, T., Kamerling, J. P., and Vliegenthart, J. F., Synthesis of hyaluronic acid-related di-, tri-, and tetra-saccharides having an N-acetylglucosamine residue at the reducing end, *Carbohydr. Res.*, 255, 61–85 (1994).

135. Slaghek, T. M., Hypponen, T. K., Ogawa, T., Kamerling, J. P., and Vliegenthart, J. F., Synthesis of hyaluronic acid related di- and tetra-saccharides having a glucuronic acid at the reducing end, *Tetrahedron. Asym.*, 5, 2291–2301 (1994).

136. Blatter, G. and Jacquinet, J. C., The use of 2-deoxy-2-trichloroacetamido-D-glucopyranose derivatives in syntheses of hyaluronic acid-related tetra-, hexa-, and octa-saccharides having a methyl beta-D-glucopyranosiduronic acid at the reducing end, *Carbohydr. Res.*, 288, 109–125 (1996).

137. Adamski-Werner, S. L., Yeung, B. K., Miller-Deist, L. A., and Petillo, P. A., Gram-scale syntheses of the (13)-linked and (14)-linked hyaluronan disaccharides, *Carbohydr. Res.*, 339, 1255–1262 (2004).

138. Plante, O. J., Palmacci, E. R., and Seeberger, P. H., Automated solid-phase synthesis of oligosaccharides, *Science*, 291, 1523–1527 (2001).

139. Palmacci, E. R., Plante, O. J., Hewitt, M. C., and Seeberger, P. H., Automated synthesis of oligosaccharides, *Helvetica Chim. Acta*, 86, 3975–3990 (2003).

140. Palmacci, E. R. and Seeberger, P. H., Toward the modular synthesis of glycosaminoglycans: synthesis of hyaluronic acid disaccharide building blocks using a periodic acid oxidation, *Tetrahedron*, 60, 7755–7766 (2004).

141. De Luca, C., Lansing, M., Martini, I., Crescenzi, F., Shen, G.-J., O'Regan, M., and Wong, C.-H. Enzymic synthesis of hyaluronic acid with regeneration of sugar nucleotides, *J. Am. Chem. Soc.*, 117, 5869–5870 (1995).

142. DeAngelis, P. L., Microbial glycosaminoglycan glycosyltransferases, *Glycobiology*, 12, 9R–16R (2002).

143. DeAngelis, P. L., Oatman, L. C., and Gay, D. F., Rapid chemoenzymatic synthesis of monodisperse hyaluronan oligosaccharides with immobilized enzyme reactors, *J. Biol. Chem.*, 278, 35199–25203 (2003).

144. DeAngelis, P. L., Hyaluronan synthases: fascinating glycosyltransferases from vertebrates, bacterial pathogens, and algal viruses, *Cell Mol. Life Sci.*, 56, 670–682 (1999).

145. Jing, W. and DeAngelis, P. L., Dissection of the two transferase activities of the *Pasteurella multocida* hyaluronan synthase: two active sites exist in one polypeptide, *Glycobiology*, 10, 883–889 (2000).

146. Balazs, E. A., Hogberg, B., and Laurent, T., Biological activity of hyaluron sulfuric acid, *Acta Physiol. Scand.*, 23, 168–178 (1951).

147. Chang, N., Intrieri, C., Mattison, J., and Armand, G., Synthetic polysulfated hyaluronic acid is a potent inhibitor for tumor necrosis factor production, *J. Leukoc. Biol.*, 55(6), 778–784 (1994).

148. Whistler, R. L. and Spencer, W., in *Sulfation by Triethylamine-Sulfur Trioxide Complex in Methods in Carbohydrate Chemistry*, Whistler, R. L., Ed., Academic Press, New York, 1964, pp. 297–298.

149. Phillips, G. B. and O'Neill, M., Sterilization (Chapter 78) in *Remington's Pharmaceutical Sciences*, Gennaro, A. R., Ed., Mack Publishing Company, Easton, Pennsylvania, 1990, pp. 1470–1480.

3

Rheological Properties of Hyaluronan as Related to Its Structure, Size, and Concentration

3.1   Some Basic Concepts of Intrinsic Viscosity

There have been extensive publications on the rheology of hyaluronic acid (HA), many of which take a complicated, theoretical approach to the subject.[1–11] The goal of this chapter, however, is to put in perspective some basic concepts of the rheology critical to the understanding of the biophysical property of HA and its derivatives, while reducing the mathematical formula and calculations to the minimum. The study of the structure–property relationship has always been an essential part of the basic science for any biomaterial, and for HA, the property that stands out is its viscosity.

3.1.1   Intrinsic Viscosity — A Measure of Hydrodynamic Volume

The viscosity of a fluid by definition is a measure of its resistance to flow.[12] Perhaps what interests us most from the chemistry standpoint is the intrinsic viscosity (IV) that largely reflects the intrinsic nature of the polymer — its molecular weight (MW), primary and secondary structures, and so on. IV $[\eta]$ is usually measured by a capillary viscometer, and a series of dilutions of the polymer are allowed to pass through the capillary under gravity.

The ratio t/t_0, the efflux time t of the solution over the efflux time t_0 of the solvent, is defined as relative viscosity η_r. Specific viscosity is defined as $\eta_{sp} = \eta_r - 1$, and reduced viscosity is given as $\eta_{red} = \eta_{sp}/C$, in which C is the concentration (g/ml). The extrapolation of reduced viscosity to zero concentration gives IV $[\eta]$. The unit of IV is milliliters per gram, which has important significance in the study of HA and its derivatives. IV is the measure of the hydrodynamic volume of polymers (Fig. 3.1).[12]

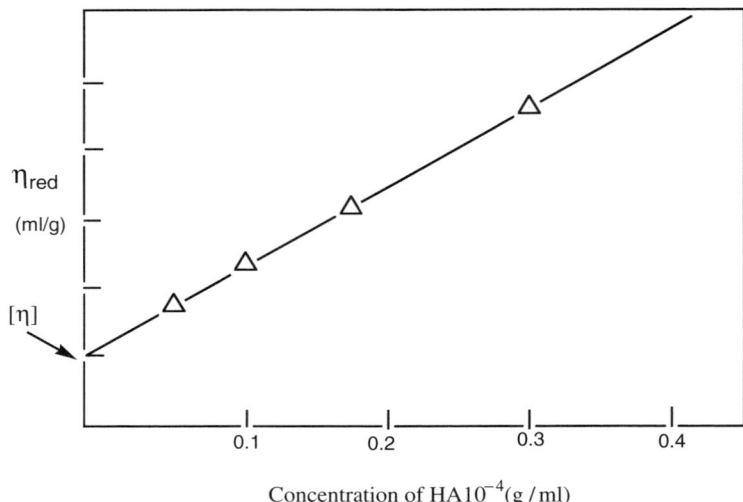

FIGURE 3.1
Intrinsic viscosity [η] is measured by extrapolating reduced viscosity to zero concentration.

3.1.2 End-to-End Distance Determines Hydrodynamic Volume

The hydrodynamic volume as a macroproperty of polymers is in essence a measure of its molecular size.[12] One way to describe the molecular size of a linear polymer is to use the end-to-end-distance theory.[13] According to that theory, the average end-to-end distance of the polymer determines the hydrodynamic volume of a linear polymer. In other words, the molecular size is conceptualized as how much space a linear polymer molecule can occupy in the solution, which is dependent on the time-average distance of the two ends of the linear polymer chain.

3.1.3 Extended Conformation Increases Molecular Size

A question of considerable interest is what factors determine the end-to-end distance of a polymer. One obvious answer is the MW. A higher MW means more atoms and therefore longer contour chain length — the sum of the lengths of all bonds. The "random-flight" model is a mathematical tool used to calculate the end-to-end distance from the bond length and bond number, assuming that the linear polymer chain had complete flexibility.[13] The equation for such imaginary conditions is:

$$\text{Distance} = \text{bond length} \times \text{bond numbers}^{1/2}$$

This random-flight model is in reality far from being a close approximation to the reality, because atoms of the polymer molecule are not in absolutely random motion (although the term "random coil" is used in its relative sense). All polymer molecules are more or less extended, and therefore the actual end-to-end distance is always larger than calculated from the random-flight model. It is obvious that an extended conformation of a polymer is favorable to the increase of molecular size, and therefore the increase of its hydrodynamic volume. This is further examined, as follows, from the structural characteristics of a molecule.

3.1.4 Structure Features Favoring Extended Conformation

Bond angles of polymer chains are fixed and always larger than 90°. Rotations of the chain bonds are somewhat restricted, even with highly flexible polymers such as polyethylene. The repulsion between adjacent hydrogen atoms makes certain more folded conformations very unfavorable.

The polysaccharide chains are linked by sugar rings. An interesting comparison was made between the IVs of the linear dextran and amylose with the same MW (1 million Da).[14-16] Under the aqueous condition, the IV of amylose was 159 ml/g, and that of dextran was 98 ml/g. The lower hydrodynamic volume of dextran is a result of the additional freedom of rotation at C5-C6 in 1,6-linked polysaccharides (Fig. 3.2).[17] For polysaccharides in which successive residues are separated by only two bonds (such as 1-4 linkages in amylose), the rotations are more restricted, and less than 5% of the conformational space is readily accessible.[18,19]

The functions and properties of polysaccharides, as well as proteins and nucleic acids, are all related to their secondary structures. The difference, however, is that polysaccharides are generally composed of repeating saccharide units that act cooperatively to form chain packing. This compensates for the entropy drive and makes the extended conformation more favorable. Chain folding within a confined volume, as seen in globular proteins, is rare for polysaccharides.[20] The secondary structure has an even more significant effect on HA, as will be discussed later.

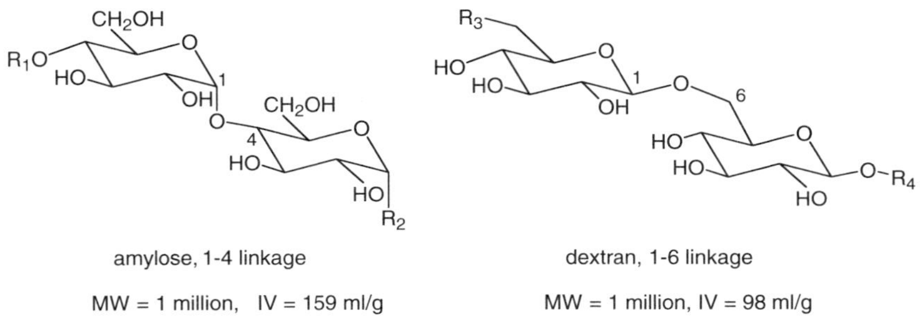

amylose, 1-4 linkage
MW = 1 million, IV = 159 ml/g

dextran, 1-6 linkage
MW = 1 million, IV = 98 ml/g

FIGURE 3.2
Comparison of hydrodynamic volume of 1 million Da molecular weight amylose and dextran.

3.1.5   Mark's Equation — IV, Rigidity, and MW

Measurement of the dilute solution viscosity of polymers has been used to study polymers since the early period of the last century, and it was Hermann Staudinger (1953 winner of Nobel Prize in Chemistry) who proposed that $\eta_{red} = KM$ (i.e., reduced viscosity is directly proportional to the polymer MW).[13]

Later, Herman Mark et al. made some modification of this hypothesis by raising the MW M to a power and established the following equation: $[\eta] = KM^{\alpha}$, where $[\eta]$ is IV, and M is MW.[13] The exponent α is a parameter that indicates the rigidity of polymer. The larger the exponent α, the more rigid the molecule is. Each polymer would have its IV $[\eta]$ that is dependent on the nature of the polymer concerned and is also affected by the temperature and the solvent in which the polymer is dissolved.

From Mark's equation, it is clear that the rigidity (extendedness) contributes exponentially to the viscosity (dynamic volume, molecular size) of the molecule. The exponent α and the factor K can be obtained by plotting the MWs and IV data of a certain polymer. The equation $\log [\eta] = \log K + \alpha \log M$ is derived by taking the logarithm of each side of Mark's equation $[\eta] = KM^{\alpha}$. The exponent α can be determined when $[\eta]$ is measured and plotted against the MW, which can be calibrated by light scattering.

3.2   Hydrodynamic Volume and Molecular Structures

3.2.1   Hydrodynamic Volume and the Structure of HA

Laurent et al.[21] established the relationship between IV and MW of HA from bovine vitreous and determined the value of K as 0.036 and that of the exponent α 0.78 when HA was dissolved in 0.2 M NaCl. Similarly, Ueno[22] studied rooster comb HA and found K as 0.039 and the exponent α as 0.77. Cleland et al.[9] used pooled HA from miscellaneous sources and determined that in 0.2 M NaCl, K = 0.0228 and α = 0.816. Cleland also found that the exponent α around 0.8 is typical for HA with MW > 10^5.

IV is often measured by researchers and manufacturers to calculate the MW of HA as well. The correlation of IV and MW is tabulated by listing the data side by side, using the K and α from the studies of Laurent,[21] Ueno,[22] and Cleland[9] (Table 3.1).

What sense can we make of those data? First, it can be found that by using the above equations, the IVs of HA turn out to be very large. For a 1 million MW of HA, the IV is over 1700 ml/g. Second, the equations show that for higher-MW HA, the exponent α (a parameter of rigidity) is very close to 0.8. According to polymer science, flexible polymers have the exponent α in the range of 0.5 to 0.8. As elements of stiffness enter the conformation, the parameter α rises to and above 0.8. This value indicated that HA is a partially

TABLE 3.1

Intrinsic Viscosity [η] Numbers and Their Corresponding Molecular
Weight M of Native HA, Based on Mark's Equation [η] = KM^α

[η] = KM^α	K = 0.036, α = 0.78 Laurent et al.	K = 0.039, α = 0.77 Ueno et al.	K = 0.0228, α = 0.816 Cleland et al.
[η] ml/g	MW 10^3 Da	MW 10^3 Da	MW 10^3 Da
1000	498	532	488
1100	562	602	549
1200	629	674	611
1300	697	748	674
1400	766	824	738
1500	837	901	803
1600	909	979	869
1700	983	1060	936
1800	1058	1141	1004
1900	1134	1224	1073
2000	1211	1309	1142
2100	1289	1394	1213
2200	1368	1481	1284
2300	1448	1569	1355
2400	1529	1658	1428
2500	1611	1749	1501
2600	1695	1840	1575
2700	1779	1932	1650
2800	1863	2026	1725
2900	1949	2120	1801
3000	2036	2216	1877
3200	2211	2410	2032
3400	2390	2607	2188
3600	2572	2808	2347
3800	2757	3012	2508
4000	2944	3219	2671
4200	3134	3430	2835
4400	3326	3644	3002
4600	3522	3860	3170
4800	3719	4080	3339
5000	3919	4302	3511

stiffened coiled molecule.[23] Considering its high MW, the long contour chain
length, and thus the tendency toward randomness, such rigidity of HA can only
be considered as being truly remarkable. When the MW is below 100,000 Da,
the HA chains become even more rigid and the value of the exponent α rises
to close to one.[24–26] Turner et al.[27] studied the viscometry of HA oligosaccharides (MW up to 30 K) and determined the exponent α to be 1.16. The higher
rigidity of HA must have its molecular basis.

For polyelectrolytes, the electrostatic repulsion of the carboxyl groups on
the polymer chain is one of the driving forces in the forming of an extended
conformation.[28] The electrostatic repulsion alone, however, does not give the
complete answer. A convincing example is the comparison of the IV data of

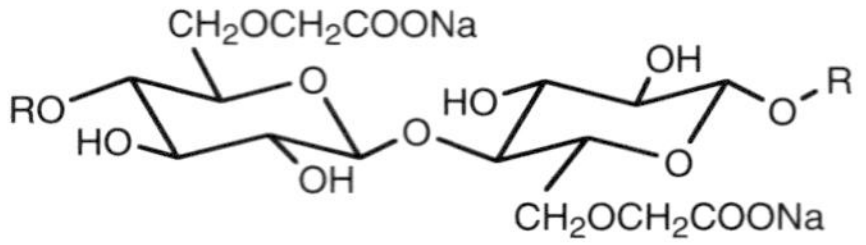

Carboxymethyl cellulose, MW = 1 million, IV= 1,180 ml/g

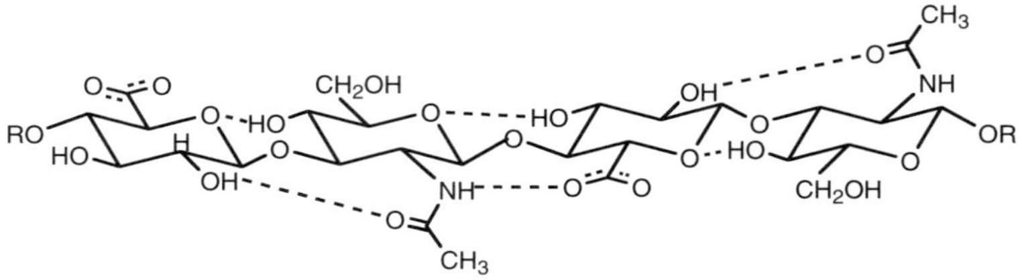

Sodium hyaluronate, MW = 1million, IV = 1,790 ml/g

FIGURE 3.3
Comparison of hydrodynamic volume of hyaluronic acid and carboxymethyl cellulose with 1 million Da molecular weight.

HA and carboxymethyl cellulose (CMC) with the same MW — 1 million Da.[14,29] The IV for HA is about 1790 ml/g, whereas that of CMC is only 1180 ml/g (Fig. 3.3). There are two factors actually more favorable to CMC than HA in terms of forming a more extended conformation. First, CMC has higher negative charge density, because CMC has one carboxylate group in every monosaccharide unit, whereas HA has one in every disaccharide unit. Second, the exclusive 1-4 linkage of CMC, other things being equal, is more likely to form a more extended conformation than the alternate 1-4, 1-3 linkage of HA. Obviously, there are some extremely important structural features in HA that can outweigh these unfavorable factors. This was recognized as related to the strong intramolecular hydrogen bonding of HA.[14]

The existence of hydrogen bonding in HA has been indicated by the nuclear magnetic resonance study, the decrease of viscosity under alkaline conditions, the slow oxidation rate in periodate treatment, and the lower pKa as compared with glucuronic acid as mentioned in Chapter 1. It was also substantiated by the x-ray diffraction analysis of the oriented HA fibers, which was found as the antiparallel double helices with axial translation of 0.85 nm per disaccharide unit (screw symmetry 4_3). This structural feature of HA is unique in glycosaminoglycans.[30] Moreover, the molecular morphology in terms of the axial periodicity per disaccharide unit was basically preserved,[23] whether HA was dehydrated or hydrated,[31-33] unprotonated or hemiprotonated.[34] The x-ray study indicated not only intrastrand hydrogen bonding but also interstrand hydrogen bonding, such as the intraduplex carboxyl–carboxylate hydrogen bonding and the hydrogen bonding with two water molecules. Furthermore, sodium ions were found nestling into the pockets of the HA chain, with each ion coordinating with six oxygen atoms in three different HA chains.[31]

Two important concepts deserve further elaboration. First, the rigidity of HA can only be conceived in a relative sense. The hydrogen bonding of HA has been described as a fleeting-association phenomenon in solutions, meaning that some segments may be more involved in hydrogen bondings at a particular moment than the others, and that the bonding may break and reform in a state of dynamic equilibrium.[20] In HA solutions, there may exist more flexible segments as well as more rigid regions, although in general, as a large macromolecule, HA still behaves like a "random coil." Second, it is necessary to distinguish the overall end-to-end distance as a macroproperty of HA from the distance held by a small segment of the molecule as a microfeature of HA. According to the x-ray analysis, the axial periodicity per disaccharide unit of HA (0.85 nm) is actually shorter than those found in other glycosaminoglycans (0.92–0.98 nm). Nevertheless, the unique potential for HA to form double helices conceivably makes the stretch of the HA molecules more stable in aqueous solutions. In other words, there might exist in the solution of HA a larger portion of extended segments than is found in the other glycosaminoglycans.

The molecular basis of large hydrodynamic volume of HA can be summarized as follows:

1. High molecular weight
2. Electrostatic repulsion of carboxylate ions
3. High restriction of rotation along chains as a two-bond linked polysaccharide
4. Intramolecular hydrogen bonding, in which all three functional groups of HA, carboxyl, hydroxyl, and N-acetyl, and water molecules are involved
5. Double-helical structure strengthened by the interstrand hydrogen bondings of the intraduplex carboxyl–carboxylate and water bridge
6. Coordination of sodium counterion with the oxygen atoms of HA

Points 2–6 all contribute to the rigidity of HA. Given the MW, they determine the actual hydrodynamic volume of HA. The same principle is used in the interpretation of the viscosity data of the carbodiimide-modified HA, as presented in the following section.

3.2.2 Hydrodynamic Volume of Carbodiimide-Modified HA

IV data for carbodiimide-modified HA derivatives (Chapter 2, Section 2.7.1), wherever applicable, are listed in Table 3.2. Almost invariably, there was considerable increase of the IV of the modified HA as compared to the native

TABLE 3.2

The Intrinsic Viscosity of Carbodiimide-Modified Hyaluronic Acid

Carbodiimides (CDI)	Degree of Coupling, %	Intrinsic Viscosity (IV) (ml/g)	IV Increase vs. Control 1, Starting HA, %	IV Increase vs. Control 2, Stirred HA, %
Mono-CDI	1.8	2866	21.8	35.2
R = octyl	4.2	2730	16.0	28.8
R' = ethyl	8.4	2534	7.6	19.5
Mono-CDI	<0.5	2401	2.0	13.3
R = tridecyl	<1.0	2591	10.1	22.2
R' = ethyl	<2.0	2651	12.6	25.0
Mono-CDI				
R = Cbz-HDA	15	2734	16.1	29.0
R' = ethyl				
bis-CDI-hexyl	3	2554	8.5	20.5
bis-CDI-phenyl	8	3073	30.5	45.5
	13	Insoluble	—	—

Notes: Cbz = carbonylbenzyloxy; HDA = hexyldiamine

HA (control 1). When HA was processed at pH 4.75 without the addition of a carbodiimide, the viscosity dropped approximately 10% (control 2). Taking this into account, the increase of the IV attributed to the carbodiimide treatment was even more significant. The degree of coupling or crosslinking was calculated as the molar percentage ratio of the consumed proton and the applied carbodiimides.

The viscosity increase varied with the degree of coupling. However, a universal pattern of direct proportion was not observed. For the octyl-ethyl carbodiimide-modified HA, the tendency was even reversed. There appeared to be a subtle effect of the size and hydrophobicity of the ligands on the secondary structures of the modified HA, and there seemed to exist an optimum degree of coupling for the viscosity increase of the modified HA, depending on the features of the carbodiimides. Whatever the rationale might be, this at minimum rules out the possibility that the viscosity increase was caused by the extra MW added to HA through a coupling reaction. Moreover, it is hardly conceivable that less than 1% of weight increase, as in the case of the coupling of octyl-ethyl carbodiimide to HA (1.8% coupling), could have caused a 35.2% increase of IV.

The crosslinking data indicated that at the low degrees of crosslinking (3% for aliphatic biscarbodiimide and 8% for aromatic biscarbodiimide), the HA molecules were still basically linear, and the MW and contour length of HA were increased mainly through the linking of HA chains by biscarbodiimides. Only in that case did the MW increase of HA become a prominent factor in enhancing its hydrodynamic volume.

On the basis of the above-mentioned data and analysis, the structure–property relationship of the carbodiimide-coupled HA is suggested as follows.

First, in the lower degrees of the carbodiimide modifications of HA (less than 16% for coupling reaction and less than 8% for crosslinking), the conformational characteristics of HA resulting from intramolecular and intrahelical hydrogen bonding appear to have been largely maintained. Alternatively, these characteristics may have been altered in a manner favorable to enhancing the rigidity and overall extendedness of HA. Second, the added side chains may have further restricted the rotation of the HA chains, thus making HA molecules further stretched. Third, the hydrophobic force caused by the hydrophobicity of the immobilized side chains in an aqueous solution might be instrumental in driving the side chain toward the inside space of the helical structure, thus further stiffening and extending the HA macromolecule.

The higher degree of crosslinking (13% for biscarbodiimide based on proton uptake) has also greatly altered the properties of HA in aqueous solution, but with a different mechanism that is forming a network. It absorbed water that amounted to 300 times the weight of crosslinked HA and became a highly swollen gel.

3.3 Absolute Viscosity Measured at Steady State

3.3.1 Networking at Higher Concentrations

As mentioned above, IV is measured at very low concentrations (such as at 0.01% w/v). Therefore, the hydrodynamic volume is considered as the property of the HA molecules and helices in their isolated states. When the concentration of HA is sufficiently high (such as 1% w/v), its network property becomes obvious. HA is completely soluble in water, but the dissolution of dry, stringy HA in water is usually rather slow. At refrigerated temperature without agitation, a high-MW solid HA (>1 million Da) could stay heterogeneous in aqueous media for days or even weeks. This indicates the strong intermolecular association of HA. At a concentration of 1% w/v, the interaction between the HA molecules is still significant. In the presence of excess water, and with intermittent and gentle swirling at room temperature, an unmodified HA viscous solution (1% 1 million MW) could fully disperse within a couple of days, whereas the EDC-modified and the ethyloctylcarbodiimide-modified HA stayed heterogeneous for double the amount of time. The networking property of HA as reflected in its gel-like property is readily observable and can be characterized and quantified by the measuring of absolute viscosity and elasticity.

3.3.2 Dynamic Viscosity

Using a Couette viscometer consisting of two concentric cylinders, with the liquid occupying the space between the adjacent surfaces, one can measure viscosity in an absolute physical sense. The ratio of the force per unit area (dyne/cm^2) to the velocity gradient (cm/second over cm = 1/second) gives units of dynamic viscosity — poise (dyne-second/cm^2). Dynamic viscosity can also be expressed as Pascal-second (Pas). One milli Pascal-second (mPas) is equal to one centipoise (cP).

Another unit closely related to dynamic viscosity and often used for HA is kinematic viscosity, which is defined as the quotient of the dynamic viscosity and the density. Because the density of HA solutions in low concentration is virtually 1, the numbers of the dynamic viscosity and kinematic viscosity of HA are practically the same. The unit of kinematic viscosity is Stoke (cm^2/second), and is often expressed as centiStoke (cSt).

Absolute viscosity of HA products has particular relevance to their use in ophthalmic surgeries.[35] To help readers put viscosity in perspective, Bothner and Wik[36] described the following reference points: "the viscosity of water is 1 mPas," "solutions with viscosity below 1000 mPas are almost watery," "a solution with a viscosity of 10,000 mPas is experienced as viscous," and "a solution with a viscosity above 100,000 mPas has a gel-like appearance."

3.3.3 Pseudoplasticity

Velocity gradient is often referred to as shear rate. HA is subjected to shear rate when going through a capillary viscometer for IV measurement, as mentioned previously.[37] In the use of absolute viscometer, shear rate is determined by the speed, shape, area, and other design characteristics of the cylinder of the viscometer. For Newtonian fluids like water and mineral oils, the viscosity is independent of the shear rate applied. An aqueous solution of HA, however, is non-Newtonian. Otherwise stated, the viscosity of HA changes as the shear rate changes. In the case of HA, as the shear rate increases, the viscosity decreases — a phenomenon called "shear-thinning," a property named pseudoplasticity (Fig. 3.4).

Pseudoplasticity is probably the most important rheological property of HA bulk solutions as far as its clinical utility is concerned, especially for its use in ophthalmic surgeries such as cataract removal and insertion of intraocular lens. Amvisc, for example, possesses very high viscosity (100 KcPs) at stationary state, which is responsible for maintaining the depth of anterior chamber. During the process of implanting an intraocular lens, the shear rate corresponds to a low range around 1/second, and Amvisc viscosity is reduced to about 40 KcPs because of the shear thinning effect. This viscosity is sufficiently high to keep HA in place but is reduced significantly to let the lens pass through — thus allowing the HA formulation to play a dual role of space filler and lubricant. When HA is expelled from the syringe through the cannula, a shear rate in the magnitude of 1000/second is generated. Because of the pseudoplastic property, HA offers little resistance when extruded through the cannula.[38]

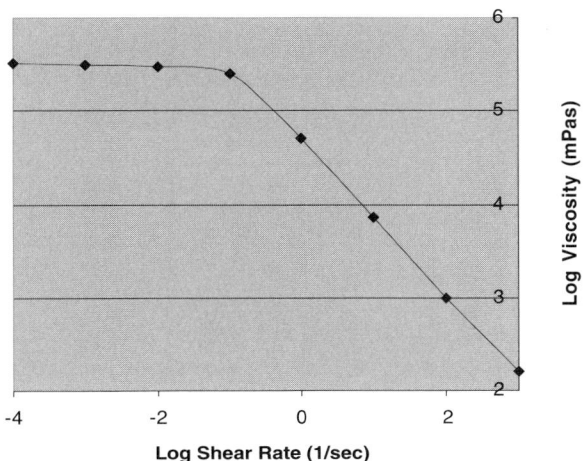

FIGURE 3.4
Pseudoplasticity or shear thinning of hyaluronic acid.

For HA, viscosity can be properly defined only in conjunction with the shear rate under which the viscosity is measured. Comparing the dynamic viscosity should always take the shear rate into consideration. For instance, 1 second^{-1} is considered a low shear rate at steady state, which is useful in comparing the viscosity of HA-based viscoelastic products for most practical purposes.[39]

3.3.4 Zero Shear Viscosity, MW, and Concentration

As shear rate approaches zero, a maximum viscosity number — zero shear viscosity — is reached. Zero shear rate viscosity has gained popularity in its use as a standard measure for the comparison of various HA products, especially those for ophthalmic use.[35] Zero shear viscosity is a function of MW and concentration. Bothner and Wik[36] studied the relation between (a) zero shear viscosity of HA, and (b) the product of HA concentration (mg/ml) × HA MW. Within the range of MW (1–4 million Da) and concentrations (10 and 20 mg/ml) investigated, there appeared to be a linear relation between log (a) and log (b). The authors concluded that a twofold increase in HA concentration or MW would result in a 10-fold increase in the zero shear viscosity (Fig. 3.5).

Alternatively, if we directly plot the zero shear rate viscosity against MW × [HA] without taking a logarithm, the viscosity appears to be a power function of MW × [HA]. It is not difficult to envision from the chart that a moderate increase of either MW or [HA] would greatly increase the zero shear viscosity (Fig. 3.6).

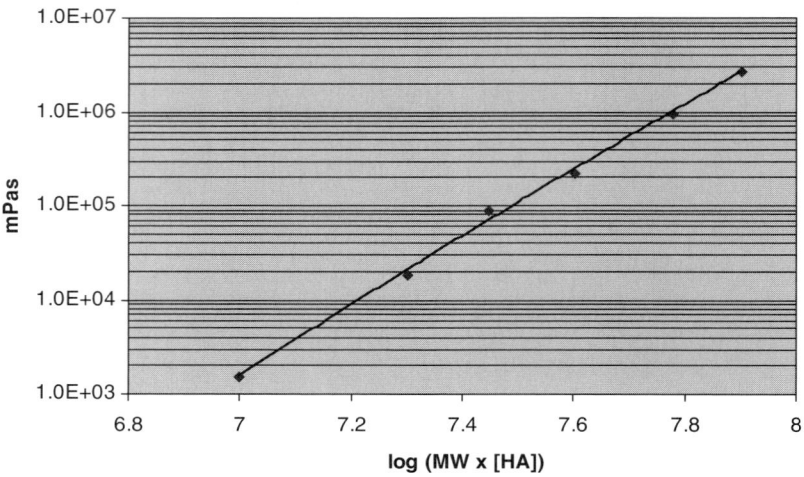

FIGURE 3.5
The logarithm of zero shear viscosity plotted against the logarithm of HA molecular weight (mw) × concentration. [HA] is HA concentration mg/ml.

3.3.5   Specific Intermolecular Interaction

Welsh et al. observed an interesting phenomenon that offered evidence that the intermolecular interaction of HA is specific, rather than random entanglement.[40] It was found in their experiments that the dynamic viscosity of HA polymer (3500 disaccharide units) was decreased by an order of magnitude

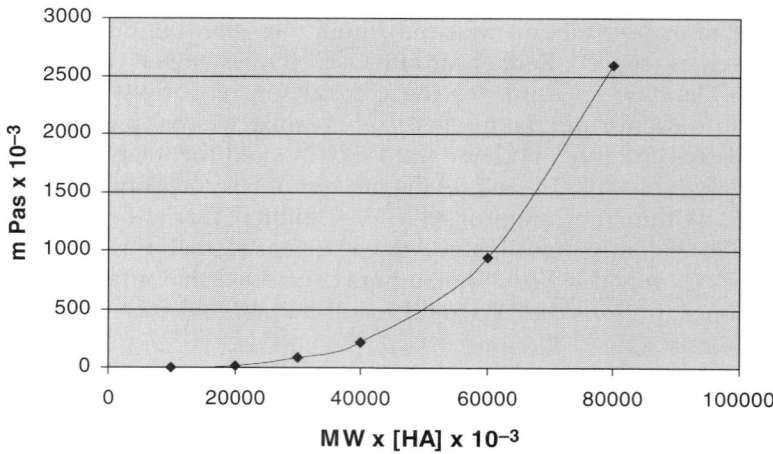

FIGURE 3.6
Zero shear viscosity of HA — a power function of HA MW × [HA].

when an equal amount of HA oligosaccharides (60 disaccharide units) was added. This is incompatible with the "entanglement coupling" theory for synthetic polymers, according to which the additional polymers should have increased the viscosity. The existence of cooperative interchain (interhelical) association and competitive inhibition of interchain (interhelical) binding of HA was therefore suggested. The HA segments were considered as occupying the binding junctions without contributing to the network.[40]

3.4   Viscoelasticity

As the average MW and the concentration of HA reach certain values, the networking property becomes obvious. This property can be quantified in at least two ways. As mentioned in the previous chapter, the bulk viscosity is normally measured at a steady shear rate. To investigate the elasticity of HA, an oscillating rheometer was used by Gibbs et al.,[11] which provides information with regard to the relative viscous property and elastic property of an HA solution. Viscosity of a real HA solution always goes hand in hand with its elasticity — the question is the relative degree of each under a given condition.

3.4.1   Elasticity Stores Energy, Viscosity Dissipates Energy

A model for an ideal elastic matter is a spring. The force F (stress) to pull the spring is directly proportional to the deformation (strain) of the spring, and stress and strain are "in phase." The energy is fully stored during the deformation, as a spring will return to its original position once it is released.

A model for purely viscous substance is a dashpot. The force to pull the dashpot is independent of the deformation (strain) but proportional to the velocity of the deformation. The energy is lost in the process, as the dashpot will not jump back to its original position when the force is released. Note that the maximum deformation occurs when the velocity reaches zero — the same time the system changes direction. Therefore, the strain and stress are "out of phase."

3.4.2   Viscoelasticity Measured under Oscillating Force

On the basis of the above principle, rheologists and engineers have devised sophisticated instruments to measure the viscoelasticity of a material or formulation. In the case of a Couette rheometer, the inner cylinder oscillates through a small angle, and the movement of HA solution transmits a sinusoidal torque to the wall of outer stationary cylinder. Both the magnitude and timing of the stress σ and the deflection γ (strain) can be measured and an equation σ = G^*. γ can be established (G^* is·a complex shear modulus of G' and G''). The in-phase component G' represents the elasticity, and the out-of-phase component G'' represents the viscosity. The unit of G is the same as σ = dyns/cm², because γ is defined as the ratio of lengths after and before deformation, and therefore unitless.

3.4.3 More Elastic and Less Viscous at Higher Frequencies

At slower oscillation, HA networking molecules would have time to disentangle; therefore, the HA solutions tend to be more viscous than at faster oscillation. At fast oscillation, the intertwined, coordinated HA molecules would not have sufficient time to disentangle, and therefore HA solutions tend to be more elastic than at slower oscillation.

As the shear rate increases, the loss modulus G'' and storage modulus G' cross over. Above the shear rate at that point, the solution is considered more elastic than viscous. DeVore and Loftus[41] compared HA solution with different MW and found that the intersection of G' and G'' lines moved from low shear rate (left) to high shear rate (right) as MW decreases (2.08×10^6, 1.57×10^6, 1.08×10^6). The corresponding kinematic viscosity values (steady state) showed a different pattern (39 KcPs, 57 KcPs, and 5 KcPs). This indicates that although both MW and concentration contribute to elasticity, MW seems to be the determining factor (Fig. 3.7).

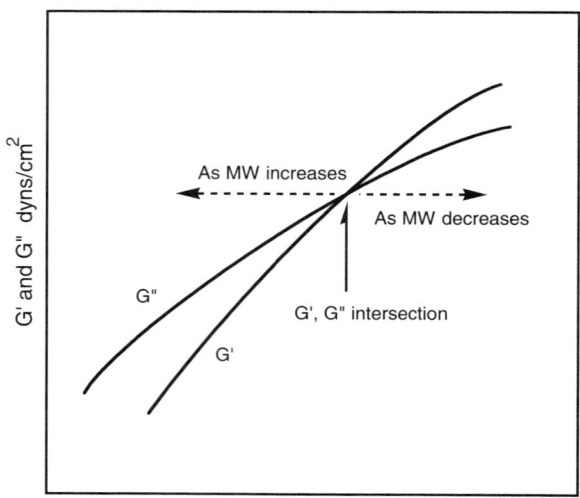

Shear rate, radians/sec

FIGURE 3.7

Dynamic viscosity sweep of HA. As shear rate increases, the elasticity G' increases more rapidly than viscosity G''. HA solution is predominantly elastic when the shear rate is above the $G'G''$ intersection.

For the biological relevance of the above phenomenon, one often-mentioned example is the relation between the viscoelasticity and the function of the joint synovial fluid. Ogston and Stanier[42,43] attributed the elasticity of bovine synovial fluid to the HA content of the fluid: "The shear rates on human joints have been estimated to be about 2.5 cycles/second during running and 0.5 cycles/second during walking," according to Balazs and Delinger.[44] It can be argued at least in rheological theory that for fast joint movement, higher MW HA preparation and crosslinked HA are more elastic than viscous, thus providing more cushioning protection to the joint than the low-MW, noncrosslinked HA preparations. The question of how significant that extra elasticity actually is in clinical practice will be discussed in Chapter 6.

3.4.4 Crosslinked HA Stays Elastic over Wide Frequencies

Wik and Wik[45] reported the data of relative elasticity and viscosity of HA and hylan gel. Hylan gel, a crosslinked HA, showed a consistently high elasticity (80–90%) over a wide frequency range (0.001–10 Hz). This phenomenon is characteristic of chemically crosslinked HA and supports the rationale rheologically for using crosslinked HA as a synovial fluid supplement.

References

1. Fouissac, E., Milas, M., Rinaudo, M., and Borsali, R., Influence of the ionic strength on the dimensions of sodium hyaluronate, *Macromolecules*, 25, 5613–5617 (1992).
2. Fouissac, E., Milas, M., and Rinaudo, M., Shear-rate, concentration, molecular weight, and temperature viscosity dependences of hyaluronate, a wormlike polyelectrolyte, *Macromolecules*, 26, 6945–6951 (1993).
3. Gomez, J. E. and Thurston, G. B., Comparisons of the oscillatory shear viscoelasticity and composition of pathological synovial fluids, *Biorheology*, 30, 409–427 (1993).
4. Mo, Y. and Nishinari, K., Rheology of hyaluronan solutions under extensional flow, *Biorheology*, 38, 379–387 (2001).
5. Yanaki, T. and Yamaguchi, T., Temporary network formation of hyaluronate under a physiological condition. 1. Molecular-weight dependence, *Biopolymers*, 30, 415–425 (1990).
6. Reed, W. F., Reed, C. E., and Byers, L. D., Random coil scission rates determined by time-dependent total intensity light scattering: hyaluronate depolymerization by hyaluronidase, *Biopolymers*, 30, 1073–1082 (1990).
7. Cleland, R. L., Ionic polysaccharides. II. Comparison of polyelectrolyte behavior of hyaluronate with that of carboxymethyl cellulose, *Biopolymers*, 6, 1519–1529 (1968).
8. Cleland, R. L., Ionic polysaccharides. IV. Free-rotation dimensions for disaccharide polymers. Comparison with experiment for hyaluronic acid, *Biopolymers*, 9, 811–824 (1970).

9. Cleland, R. L., Viscometric study of the proteoglycan–hyaluronate (2:1) "dimer": minimum hyaluronate chain length, *Biopolymers*, 22, 2501–2056 (1983).

10. Cleland, R. L. and Wang, J. L., Ionic polysaccharides. 3. Dilute solution properties of hyaluronic acid fractions, *Biopolymers*, 9, 799–810 (1970).

11. Gibbs, D. A., Merrill, E. W., Smith, K. A., and Balazs, E., Rheology of hyaluronic acid, *Biopolymers*, 6, 777–791 (1968).

12. Kinsinger, J. B., in *Encyclopedia of Polymer Science and Technology*, Mark, H. F. G., and Gaylord, N. G., Eds., Interscience, New York, 1971, Vol. 14, pp. 717–740.

13. Billmeyer Jr., F. W., *Synthetic Polymers — Building the Giant Molecule*, Doubleday, Garden City, NY, 1972.

14. Morris, E. R., Rees, D. A., and Welsh, E. J., Conformation and dynamic interactions in hyaluronate solutions, *J. Mol. Biol.*, 138, 383–400 (1980).

15. Kurata, M., Tsunashima, Y., Iwama, M., and Kamada, K., in *Polymer Handbook*, Brandup, J., Immergutt, E. H., and McDowell, W., Eds., Wiley, New York, 1975, Vol. 4, pp. 30–33.

16. Goodall, D. M., Polysaccharide conformations and kinetics, *Acc. Chem. Res.*, 20, 59–65 (1987).

17. Rees, D. A., Polysaccharide conformation (Chapter 8), in *Carbohydrates, MTP Int. Rew. Sci., Organic Chemistry, Series One*, Aspinall, G. O., Ed., Butterworths, London, 1973, Vol. 7, pp. 251–283.

18. Rees, D. A., Polysaccharides: conformational properties in solution, 26.4, in *Comprehensive Organic Chemistry*, Barton, D. and Ollis, W. D., Eds., Fergamon, Oxford, 1979, Vol. 5, pp. 817–830.

19. Rees, D. A. and Scott, W. E., Polysaccharide conformation. Part VI. Computer model-building for linear and branched pyranoglycans. Correlations with biological function. Preliminary assessment of inter-residue forces in aqueous solution. Further interpretation of optical rotation in terms of chain conformation, *J. Chem. Soc. (B)*, 469–479 (1971).

20. Rees, D. A. and Welsh, E. J., Secondary and tertiary structure of polysaccharides in solutions and gels, *Angew. Chem. Int. Ed. Engl.*, 16, 214–224 (1977).

21. Laurent, T. C., Ryan, M., and Pletruszkiewicz, A. Fractionation of hyaluronic acid, The polydispersity of hyaluronic acid from the bovine vitreous body. *Biochim. Biophys. Acta*, 42, 476–485 (1960).

22. Ueno, Y., Tanaka, Y., Horie, K., and Tokuyasu, K., Low-angle laser light scattering measurements on highly purified sodium hyaluronate from rooster comb, *Chem. Pharm. Bull.*, 12, 4971–4975 (1988).

23. Phelps, C. F., The dilute solution properties of glycosaminoglycans and proteoglycans, (Chapter 2), in *Molecular Biophysics of the Extracellular Matrix*, Arnott, S., Rees, D. A., and Morris, E. R., Eds., Humana, Clifton, NJ, 1983, pp. 21–39.

24. Cleland, R. L., Viscometry and sedimentation equilibrium of partially hydrolyzed hyaluronate: comparison with theoretical models of wormlike chains, *Biopolymers*, 23, 647–666 (1984).

25. Shimada, E. and Matsumura, G., Viscosity and molecular weight of hyaluronic acid, *J. Biochem.*, 78, 513 (1975).

26. Bothner, H., Waaler, T., and Wik, O., Limiting viscosity number and weight average molecular weight of hyaluronate samples produced by heat degradation, *Int. J. Biol. Macromol.*, 10, 287–291 (1988).

27. Turner, R. E., Lin, P., and Cowman, M. K., Self-association of hyaluronate segments in aqueous NaCl solution, *Arch. Biochem. Biophys.*, 265, 484–495 (1988).

28. Molyneux, P., in *Water-Soluble Synthetic Polymers: Properties and Behavior,* CRC Press, Boca Raton, FL, 1985.

29. Morris, E. R., Rees, D. A., Welsh, E. J., Dunfield, L. G., and Whittington, S. G., Relation between primary structure and chain flexibility of random coil polysaccharides: calculation and experiment for a range of model carrageenans, *J. Chem. Soc. Trans. Perkin II,* 7, 793–800 (1978).

30. Arnott, S. and Mitra, A. K., X-ray diffraction analyses of glycosaminoglycans (Chapter 3), in *Molecular Biophysics of the Extracellular Matrix,* Arnott, S., Rees, D. A., and Morris, E. R., Eds., Humana, Clifton, NJ, 1983, pp. 41–67.

31. Guss, J. M., Hukins, D. W., Smith, P. J., Winter, W. T., and Arnott, S., Hyaluronic acid: molecular conformations and interactions in two sodium salts, *J. Mol. Biol.,* 95, 359–384 (1975).

32. Mitra, A. K., Raghunathan, S., Sheehan, J. K., and Arnott, S., Hyaluronic acid: molecular conformations and interactions in the orthorhombic and tetragonal forms containing sinuous chains, *J. Mol. Biol.,* 169, 829–859 (1983).

33. Mitra, A. K., Arnott, S., and Sheehan, J. K., Hyaluronic acid: molecular conformation and interactions in the tetragonal form of the potassium salt containing extended chains, *J. Mol. Biol.,* 169, 813–827 (1983).

34. Sheehan, J. K., Gardner, K. H., and Atkins, E. D., Hyaluronic acid: a double-helical structure in the presence of potassium at low pH and found also with the cations ammonium, rubidium, and caesium, *J. Mol. Biol.,* 117, 113–135 (1977).

35. Arshinoff, S. A. and Hofman, I., Prospective, randomized trial comparing Micro Visc Plus and Healon GV in routine phacoemulsification, *J. Cataract Refract. Surg.,* 24, 814–820 (1998).

36. Bothner, H. and Wik, O., Rheology of hyaluronate, *Acta Otolaryngol. Suppl.,* 442, 25–30 (1987).

37. Yanaki, T. and Yamaguchi, M., Shear-rate dependence of the intrinsic viscosity of sodium hyaluronate in 0.2 M sodium chloride solution, *Chem. Pharm. Bull.,* 42, 1651–1654 (1994).

38. Arshinoff, S. A., Dispersive and cohesive viscoelastic materials in phacoemulsification revisited 1998, *Ophthal. Pract.,* 16, 24–32 (1998).

39. Leith, M. M., Loftus, S. A., Kuo, J. W., DeVore, D. P., and Keates, E. U., Comparison of the properties of AMVISC and Healon, *J. Cataract Refract. Surg.,* 13, 534–536 (1987).

40. Welsh, E. J., Rees, D. A., Morris, E. R., and Madden, J. K., Competitive inhibition evidence for specific intermolecular interactions in hyaluronate solutions, *J. Mol. Biol.,* 138, 375–382 (1980).

41. DeVore, D. P. and Loftus, S. A., Rheology of sodium hyaluronate solutions and relevance to their use as medical implant, *Mat. Res. Soc. Symp. Proc.,* 110, 455–461 (1989).

42. Ogston, A. G. and Stanier, J. E., Elastic properties of ox synovial fluid, *Nature,* 165, 571 (1950).

43. Ogston, A. G. and Stanier, J. E., The physiological function of hyaluronic acid in synovial fluid: viscous, elastic, and lubricant properties, *J. Physiol.,* 119, 244–252 (1953).

44. Balazs, E. A. and Delinger, J. L., Sodium hyaluronate and joint function, *J. Equine Vet. Sci.,* 5, 217–228 (1985).

45. Wik, H. B. and Wik, O., Rheology of hyaluronan, in *The Chemistry, Biology, and Medical Applications of Hyaluronan and Its Derivatives,* Portland, London, 1998, pp. 25–32.

4

Standards, Tests, and Analytical Methods

4.1 Standards and Specifications

4.1.1 European Pharmacopoeia

As published in Supplement 2000 of European Pharmacopoeia (EP), <EP 2000:1472> stipulates the key parameters of sodium hyaluronate to be analyzed and the corresponding methods to perform the analysis.[1] An updated version of the monograph of sodium hyaluronate was published in the 2001 Supplement (EP, 2001 Supplement, pp. 1415–1418).

EP was founded by eight European states (Belgium, France, Germany, Italy, Luxembourg, Netherlands, Switzerland, and United Kingdom) in 1964 and now expanded to 27 states. EP fulfills the task of establishing common standards for the analysis of the composition of substances used in medicines, so as to ensure their quality. The EP monographs replaced the old national pharmacopoeias and have the force of law. EP (http://www. pheur.org) is published by the European Directorate for the Quality of Medicines of the Council of Europe. EP contains standardized specifications defining the quality of pharmaceutical preparations and their constituents and containers.

<EP 2000:1472> provides the definition of hyaluronic acid (HA), indicates the source of HA and main characters of dry HA (powder or fibrous aggregate), and specifies the identification methods. Thereafter, the monograph describes each listed test including pH, intrinsic viscosity, sulfated glycosaminoglycans, nucleic acids, protein, chlorides, iron, loss on drying, microbial contamination, sterility, and bacterial endotoxins. A detailed description of two assays pertaining to the analysis of HA, carbazole assay for HA content and intrinsic viscosity assay related to HA molecular weight (MW), is also provided.

Table 4.1 lists comparison of most of the testing items described in EP 2000:1472 and in *ISO 15798 Ophthalmic Implants — Ophthalmic Viscosurgical Devices* (OVD), which will be described in the next section. The table is intended as an introduction and for the convenience of the readers, who are advised to refer to the original documents for the completeness and accuracy of the contents.

TABLE 4.1

Tests in European Pharmacopoeia and *ISO 15798 Ophthalmic Implants —
Ophthalmic Viscosurgical Devices (OVD)* Applicable to Hyaluronic
Acid Testing

Product Parameters	European Pharmacopoeia Solid Sodium Hyaluronate	ISO 15798 Ophthalmic Viscosurgical Devices
Appearance, absorbance, transmittance	Clear, absorbance ≤ 0.01 at 600 nm at 0.10 g/30 ml	Spectral transmittance recorded over 200–1200 nm
Identification	IR absorption and Na reaction	Chemical description required
NaHA Concentration	Carbazole assay	Description of assay for concentration
Intrinsic Viscosity and MW	IV measured by capillary viscometer (MW can be derived from IV)	Testing MW and its distribution by standard methods
Kinematic Viscosity	Not applicable	At defined shear rate between 1000 and 0.001 sec^{-1}
Elasticity	Not applicable	At frequencies of 0.01–20 Hz
pH	5.0–8.5 at 5 mg/ml	6.8–7.6
Nucleic Acids	Absorbance ≤ 0.5 (3.33 mg/ml at 260 nm)	Quantified as impurities
Protein	≤ 0.3% dry weight of NaHA	Quantified as impurities
Sulfated GAGs	Perchloric acid/$BaCl_2$ assay if the source is cock's combs	Not specified, but may fall into category of impurities
Iron	≤80 ppm of dry weight, by atomic absorption spectrometry	Not specified, but may fall into category of impurities
Microbial Count	≤10^2 microorganisms/g	Specified as bio-contaminants
Sterility	Required for sterile dosage	Contamination rate limit 10^{-3}
Bacterial Endotoxin	Endotoxin level ≤ 0.05 IU/mg, for ocular or intraarticular use	Endotoxin level ≤ 0.5 EU/ml
Osmolality	Not required	200–400 mOsm kg, cryoscopic vapor pressure osmometer
Particulates	Not required	USP 23 <788> "Particulate Matter in Injections"
Refractive Index	Not required	RI (air/OVD) at 25°C

4.1.2   Viscoelastics in Ophthalmic Application

An international standard has been established pertaining to the materials
applied in viscosurgery of the eye.[2] In 2001, International Standards Organization (ISO) published *ISO 15798 Ophthalmic Implants — Ophthalmic Viscosurgical Devices.* ISO prepares international standards through the work

of expert technical committees comprising delegates from participating countries.

The OVD project began as an initiative of the American Academy of Ophthalmology and the American National Standards Institute under the chairmanship of John Alpar. The establishment of ISO 15798, a standard for the safety, performance, and approval requirements for viscoelastics, was the result of a joint effort of many distinguished scientists and physicians from more than 20 countries.

Among all the medical applications of HA-based products, the longest history of commercialization has been in the field of ophthalmic surgery. It did not come as a surprise that the first international standard applicable to a specific use of HA was established precisely in the field of ophthalmology. The ISO standard for OVD pertains to all viscoelastic products indicated for viscosurgery of the eye. However, most of the OVD are HA based, according to Steve Arshinoff.[3] Many studies in ISO designated tests of HA based viscoelastics, such as osmolality,[4] endotoxin,[5] and effect of its use on intraocular pressure,[6] have been reported.

Most of the ISO OVD tests pertaining to HA are listed in Table 4.1, along with EP tests for comparison. Wherever available, the specifications are indicated. Readers are advised to refer to the original ISO document for complete information. Most of the information provided in the table is self-explanatory. A further study of related references, however, is recommended for a better understanding of the technical background of those methods.

4.1.3   ISO 10993 — Biocompatibility

HA-based products are regulated as medical devices for certain indications in the U.S. and many other countries, although HA in its aqueous form, or solid form, has little resemblance to a device in its traditional sense — a piece of equipment.

Biological evaluation of medical devices is to determine the potential toxicity resulting from contact of the component materials with the body. In 1986, the U.S. Food and Drug Administration (FDA), Health and Welfare Canada, and Health and Social Services UK issued the Tripartite Biocompatibility Guidance for Medical Devices. Since then, ISO developed ISO 10993 in an effort to harmonize biocompatibility testing. In 1995, the Office of Device Evaluation of FDA issued guidance memorandum G95-1, entitled *Use of International Standard ISO-10993, Biological Evaluation of Medical Devices Part-1: Evaluation and Testing,* also referred to as "Blue Book Memos," to implement ISO 10993, with certain modifications. The memo replaced the use of Office of Device Evaluation General Program Memorandum G87-1, entitled *Tripartite Biocompatibility Guidance,* dated April 24, 1987.

ISO 10993 is a series of standards, most of which are relevant to the testing of the biocompatability of HA-based material (Table 4.2). However, the standards are basically dealing with solid biomaterials composed, for example, of ceramics, plastics, or metals. Therefore, certain adjustments are necessary

TABLE 4.2

ISO 10993 Biological Evaluation of HA-Based Medical Products

ISO Number	Title
ISO 10993-1	Evaluation and testing
ISO 10993-2	Animal welfare requirement
ISO 10993-3	Tests for genotoxicity, carcinogenicity, and reproductive toxicity
ISO 10993-4	Selection of tests for interaction with blood
ISO 10993-5	Tests for *in vitro* cytotoxicity
ISO 10993-6	Tests for local effects after implantation
ISO 10993-7	Ethylene oxide sterilization residuals
ISO 10993-8	Selection and qualification of reference material for biological tests
ISO 10993-9	Framework for identification and quantification of potential degradation products
ISO 10993-10	Tests for irritation and delayed-type hypersensitivity
ISO 10993-11	Tests for systemic toxicity
ISO 10993-12	Sample preparation and reference materials
ISO 10993-13	Identification and quantification of degradation products from polymeric medical devices
ISO 10993-16	Toxicokinetic study design for degradation products and leachables from medical devices
ISO 10993-17	Methods for the establishment of allowable limits for leachable substances

For updated ISO 10993, refer to the ISO Web site (www.iso.org) or the AAMI Web site (www.aami.org).

Other ISO 10993 series standards that may not be applicable to hyaluronic acid–based products include *ISO 10993-14, Identification and Quantification of Degradation Products from Ceramics* and *ISO 10993-15 Identification and Quantification of Degradation Products from Metals and Alloys.*

for HA-based products, which in many cases are in the form of aqueous solution or suspension. For example, leachable substances may not apply to HA products per se but may be relevant to its container, such as syringes. The blood compatibility may not apply to ceramics or metals, but it becomes relevant to HA because it is miscible with blood.

ISO 10993-1, "Evaluation and Testing," uses a tabular format (matrix) for laying out the test requirements. Table 4.3 gives a list of tests that are applicable to many HA-based products within the requirement of ISO 10993 and the guidelines of the FDA "Blue Book Memos." Examples of those tests are the guinea pig maximization method[7] for a sensitization study, the Ames mutagenicity assay,[8] the chromasol aberration assay,[9] and the sister chromatid exchange assay[9] for genotoxicity (mutagenicity).

According to ISO 10993-1, the risk of a device, and accordingly the need for specific testing, is largely determined by the type of the contact tissue and residence time of the device. Take HA products for knee injection as

TABLE 4.3

Biocompatibility Tests in ISO 10993 and Related U.S. Food and Drug Administration Guidelines (1995)

Relevant Biocompatibility Tests[a]	ISO 10993[b]	FDA Guideline	Examples of Applicable Methods
Cytotoxicity	√	√	Agar Diffusion Test as in USP <87> (see Table 4.5) and USP <1031> (see Table 4.4)
Sensitization Study	√	√	Tests delayed-type hypersensitivity (e.g., guinea pig maximization method) as part of ISO 10993-10
Irritation or Intracutaneous Toxicity	√		Tests for irritation (e.g., rabbit skin tests) as part of ISO 10993-10
Acute Systemic Toxicity, <24 hours (including pyrogenicity per ISO 10993-11)	√		ISO 10993-11 references ASTM F750: <Standard Practice Evaluation Material Extracts by Systemic Injection in Mice (Method B), Intraperitoneal.>
Subchronic toxicity, up to 90 days		√	"For a material, the test is equivalent to implantation test"
Genotoxicity (mutagenicity)	√	√	Tests for genotoxicity, carcinogenicity, and reproductive toxicity, (ISO 10993-3) also Ames Mutagenicity Assay, Chromasol Aberration Assay, Sister Chromatid Exchange Assay
Implantation Test	√	√	ISO 10993-6 (Tests for Local Effects after Implantation) and USP <88> (see Table 4.5)

a Hemocompatibility, though not generally required for implantable medical device, is often examined for hyaluronic acid–containing products, because of their form as an injectable gel. *ISO 10993-4 Selection of Tests for Interactions with Blood (Thrombosis Coagulation, Platelet, Hematology)* may apply.

b All 7 biocompatibility tests are now included in ISO 10993–1 (2003) as "evaluation tests for consideration."

an example. They may be considered implant devices, having contact with "bone/tissue" when in use, and as belonging to the category of having "prolonged exposure" (<30 days), because by 30 days, over 99% of HA injected into the knee has been cleared. The higher the risk, the more tests are required. Whether each test listed is necessary, the manufacturer needs to discuss with the regulatory authorities concerned, taking into consideration many related factors, such as the general knowledge of the biological properties of HA, actual indication for use, source, and specification of a particular product, and so on.

It is desirable that most of these tests be conducted during the product development stage to demonstrate the safety profile. Some tests may not be necessary for routine control of manufacturing processes or for the release of the products. Individual manufacturers should validate the use of a battery of chemical, physical, and biological tests similar to the EP methods to satisfy the need for routine quality control.

4.1.4   U.S. Pharmacopoeia

Unlike EP, U.S. Pharmacopoeia (USP) has yet to establish the monograph of hyaluronic acid. Hyaluronate is mentioned in USP 24/NF 19 (2001 through Supplement Four) only within the content of the hyaluronidase monograph. However, many of the test methods that may be applicable to HA products, especially the bioassays mentioned in EP, ISO-OVD, and ISO 10993, can be found in USP (Table 4.4). Of course, the actual methods used for an individual product may vary, and such information is not generally available to the public.

As a participant in the International Conference on Harmonization effort, USP is expanding its toxicity procedures <1031> to be more in line with ISO 10993. Table 4.5 summarizes the status of the USP procedures as of its 2001 edition (USP 24, NF 19; it remained unchanged as of version USP 27/2004).

TABLE 4.4

USP Methods Applicable to HA-Based Medical Products

Assay Category	USP Methods
Microbial Count	<61> Microbial Limit Tests
Sterility	<71> Sterility Tests
Bacterial Endotoxin	<85> Bacterial Endotoxin Test
Pyrogen	<151> Pyrogen Test
Iron	<241> Iron
Osmotic Pressure	<785> Osmolarity
Particulates	<788> Particulate Matter in Injections
Biocompatibility	<1031> The Biocompatibility of Materials Used in Drug Containers, Medical Devices, and Implants

TABLE 4.5

Status of Biocompatibility Methods in the USP General Chapters

Biological Effect	Procedures in USP General Chapter
Cytotoxicity	*Biological Reactivity Tests, In Vitro* <87>
Irritation or intracutaneous reactivity	*Biological Reactivity Tests, In Vivo* <88>
Systemic toxicity (acute toxicity)	*Biological Reactivity Tests, In Vivo* <88>
Implantation	*Biological Reactivity Tests, In Vivo* <88>
Hemocompatibility	Under development
Sensitization	Not established
Subchronic toxicity (subacute toxicity)	Not established
Genotoxicity	Not established
Chronic toxicity	Not established
Carcinogenicity	Not established
Reproductive or developmental toxicity	Not established
Biodegradation	Not established

4.2 Analyses of HA Structure, Size, and Quantity

The analytical methods pertaining to the structure, molecular size, and quantity of HA are reviewed in this section. Readers may refer to Chapters 1 and 2 for additional information regarding the principle and mechanism of many of the chemical methods. Viscometry has been reviewed in detail in Chapter 3. However, it is worth repeating here the significance of the intrinsic viscosity of HA as a measure of its MW and the elasticity as related to the networking of HA. In a sense, intrinsic viscosity reflects the size of a single molecule, whereas elasticity relates to the interaction of a large amount of HA molecules. As the HA concentration increases, and the network characteristics of HA become significant, so does its elasticity.

4.2.1 Infrared Spectroscopy

Infrared (IR) analysis has been listed in EP as an identification method for sodium hyaluronate. The methods and the reference spectrum of sodium hyaluronate are cited (Section 2.2.24). In 1954, Orr[10] used conventional IR spectroscopy and gave partial assignments of bands of hyaluronic acid and hyaluronate (Table 4.6). HA films were cast on glass plate and then stripped off for spectral examination. Neutral solution of HA was almost fully ionized, but HA in HCl was mostly in its free acid form.

In 1976, Cael et al. reported the polarized Fourier Transform Infrared Spectroscopy (FTIR) spectra of oriented films of NaHA.[11] As compared to dispersive instruments, FT instruments have large energy output, better signal-to-noise ratio, and consequently, the capability for rapid scanning and accurate frequency determination. Polarized IR, in addition, can provide information with

TABLE 4.6

Partial Assignments of Infrared Bands of Hyaluronic Acid

Absorption	Assignment
1736 cm^{-1} (main), 1230 cm^{-1} (minor)	Free carboxylic, COOH
1610 cm^{-1} (main), 1410 cm^{-1} (minor)	Carboxylate, COO
1648 cm^{-1}, 1560 cm^{-1}, 1315 cm^{-1}	Monosubstituted amide
1375 cm^{-1} (sharp)	Associated methyl group
1000–1200 cm^{-1}	C-O stretching and C-O-H bending

regard to the molecular orientation of polymer fibers. The oriented films were prepared according to the method described by Atkins et al., in which the specimen strips were stacked and stretched under constant load at a relative humidity in the range of 75–85%.[12] The dichroism data for the vibrational modes of the amide and carbonyl groups were interpreted and found to be in agreement with contracted fourfold helical conformation as determined by x-ray.[13,14] The IR assignment of HA by Cael, though an improvement over Orr, was still partial and from solid samples only. Gilli used the FTIR instrument with attenuated total reflectance technique to analyze HA samples in aqueous solution and provided a more detailed assignment.[15] The aqueous solution spectra have the benefit of avoiding the residual cristallinity effect of "shift and splitting of frequencies arising from interactions with the crystalline lattice."

4.2.2   UV-Visible Spectroscopy

Pure hyaluronic acid is colorless. The visible absorption spectroscopic data, however, can be informative when distinct colors are developed in certain colorimetric assays, which will be the focus of this section. UV absorption of HA is quite strong at around 220 nm and therefore can be used for quantitative assays of pure HA. The UV absorbance at 232 nm is often indicative of the formation of an α, β unsaturated carboxylate moiety due to β-elimination, which has been mentioned earlier. The microparticle photometric agglutination (MPA) assay is also included in this section, but note that the measured transmittance of 800 nm only sits at the borderline of the visible region.

4.2.2.1   Carbazole Assay

Carbazole assay is one of the most commonly used assays for hyaluronic acid and is specified in EP as the assay for HA concentration. Chapter 2 has discussed the chemistry related to the carbazole assay. As an improvement to the assay, Bitter and Muir[16] (1962) used borate that enhances the red color formation and stability and decreased the sensitivity to salts. The result showed a better correlation between uronic acid and hexosamine contents. The method has become a standard method for routine quantitative determination of uronic acid. A solution containing 1 µg/ml of uronic acid can show a positive color reaction. The absorbance at 530 nm is a linear function of concentration between 4 and 40 μg/ml. It is being used mostly as a quantitative assay. At the same time, it is also qualitative, in part, that the distinct red color generated during the reaction indicates the existence of a class of carbohydrates — uronic acid.

4.2.2.2   Alcian Blue Assay

Though widely used, carbazole assay is not particularly convenient, as it requires handling of concentrated sulfuric acid and heating of the test solutions twice during the procedure. Gold[17] developed a spectrometric method using Alcian blue, which allows simple, rapid, and reliable estimation of micrograms of glycosaminoglycans including HA. HA forms a soluble complex with Alcian blue. The procedure is based on the difference of the absorbance minimum at 480 nm between the HA–dye complex (higher) and the dye control (lower). Both groups showed linear increases over the range of 5–75 μg. Albumin and individual sugars do not interfere with the assay, but high concentration of chloride ions do.

4.2.2.3   Morgan-Elson Assay

A popular assay presently used for quantifying N-acetylglucosamine content of HA and often referred to as the Morgan-Elson assay[18,19] is, in fact, a modification of the assay by Reissig.[20] In both the original version and its Reissig modification, hexosamine and N-acetyl hexosamine, after undergoing certain chemical treatment, react with DMAB to form a distinct color compound. Reissig method was characterized by substituting the borate buffer for the carbonate buffer used in Aminoff's method,[19] and as a result, the color yield increased by twofold. With the Reissig procedure, 3×10^{-10} mole of N-acetylglucosamine can be determined. The chemistry and mechanism of those reactions have been discussed in Chapter 2.

Because the Reissig method works on the reducing end of the N-acetyl hexosamine, hyaluronidase is often used to degrade HA into its disaccharide units. This makes the Reissig method more specific to HA than, for example, the carbazole method mentioned above.

Improvement on the Morgan-Elson method continued after the work of Reissig. Recently, Asteriou et al.[21] proposed a new interpretation of the Reissig signal, consisting of two factors: turbidity resulting from the formation of carbohydrate–protein complexes, and color with maximal intensity at 585 nm, resulting from the reaction of DMAB with the carbohydrate. The colorimetric intensity of HA can be estimated by subtracting the turbidity from the absorbance. The method helps to interpret the results of Reissig method, taking into account the interference from the presence of protein.

4.2.2.4   Thiobarbituric Assay

The thiobarbituric assay, developed by Jourdian et al.,[22] is a coupled enzyme-colorimetric assay, suitable for measuring HA content in a mixture of glycosaminoglycans or crude biological extracts, because of its specificity to HA.

The thiobarbiturate assay comprises the following steps: the use a purified form of a hyaluronate lyase, *Streptomyces hyalurolyticus,* which is less likely to be inhibited by other glycosaminoglycans, and is therefore more effective at degrading HA to its unsaturated disaccharide units; the oxidation of the disaccharide by periodic acid to formyl pyruvate; and its reaction with thiobarbiturate to form a chromophore with maximum absorbance at 549 nm.[23]

The Jourdian method has a linear response range of 5–100 µg. The sensitivity for HA was further improved by Adams et al. through an automatic version of the periodate-thiobarbituric acid assay, which demonstrated a linear response range between 0.1 and 2.5 µg uronic acid.[24]

4.2.2.5 Enzyme-Linked HA-Binding Assays

The assays based on specific affinity of HA to HABP (HA binding proteins) are hundreds of times more sensitive than the types of assays mentioned above. Proteins with high affinity to HA are proteoglycans that can be isolated from bovine articular cartilage,[25] bovine nasal cartilage,[26] or glycoproteins such as hyaluronectin isolated from human brains.[27] Protein binding assays can determine nanogram amounts of HA. Some HA-binding assays measure colors generated from peroxidase[25,26,28] and alkaline phosphatase reactions,[27] and therefore they also belong to the category of colorimetric assays. Refer to the review entitled "Seven Different Assays of Hyaluronan Compared for Clinical Utility," by Lindqvist et al.[29]

The Corgenix HA test kit is an enzyme-linked binding protein assay based on the work of Chichibu.[30,31] In the test, HA is sandwiched between the HABP-coated surface of microwells and HABP conjugated with horseradish peroxidase. A chromogenic substrate of tetramethylbenzidine and hydrogen peroxide is added to develop a color reaction. The optical density of the color is measured with a spectrophotometer at 450 nm. The Corgenix HA assay was intended as a diagnostic aid in the serologic evaluation of rheumatoid arthritis and liver fibrosis and cirrhosis, and it is available in the U.S. for investigational use only (www.corgenix.com). The assay has been validated, and its key performance characteristics as an HA assay are summarized in Table 4.7. The detection limit was 10 ng/ml. The accuracy (recovery)

TABLE 4.7

Enzyme-Linked Binding Protein Assay for Hyaluronic Acid by Corgenix

Performance Category	Range, Mean Values
HA assay range	0–75 ng/ml
Detection limit	10 ng/ml
Precision	Coefficient of variations (CV%)
Intraassay (1 assay, 16 replicates)	4.2%
Interassay (3 assays, 16 replicates)	6.3%
Accuracy (mean recovery %)	100.9
Linearity (r^2)	≥0.999

was determined by comparison with series of samples with known value. The precision is determined customarily by coefficient of variations (i.e., standard deviation divided by arithmetic mean).

HA-binding assays are a viable alternative to immunogenic assays that would be difficult to introduce to HA because native HA is essentially nonimmunogenic. The two types of assays have the similar level of sensitivity and operating principles.[29]

4.2.2.6 Microparticle Photometric Agglutination Assay

The microparticle photometric agglutination (MPA) assay has been developed recently by Takei et al.[32] The MPA method is also based on the interaction between HA and HA binding protein. There are a few steps in the procedure. First, HABP is covalently sensitized on the surface of microparticles; second, HA molecules adhere to the HABP, causing particles to agglutinate; third, the transmittance change at 800 nm resulting from the agglutination of the microparticles is measured; and fourth, HA concentration is calculated as a function of the magnitude of agglutination.

The lower limit of detection was 10 ng/ml. The analytical range of the assay for HA was found to be 10–1200 ng/ml. The intraassay precision (coefficient of variations) was 3.0–8.4% ($n = 10$), and the interassay precision was 4.8–8.9% ($n = 3$). The MPA method was further compared with an ELISA assay of Chugai Pharmaceuticals, Japan. The correlation coefficient of data derived from MPA method against ELISA plate method was 0.989 with a slope of 1.01 ($n = 31$).[32]

4.2.3 Nucleic Magnetic Resonance Spectroscopy

Nucleic magnetic resonance (NMR) spectroscopy is a sophisticated tool in structural chemistry. It has proved to be a valuable addition to the traditional chemical and enzymologic approaches to the structural study of HA. Three commonly used nuclei are ^{1}H, ^{13}C, and ^{15}N. NMR provides more information for structural details than IR, ultraviolet, or visible spectrometry. For every nonequivalent atom, a unique signal of NMR can be generated. Each signal has three major parameters: chemical shift, coupling constant, and relaxation times, which afford ample information for the elucidation of the primary and secondary structures of hyaluronic acid and its derivatives. The successful application of NMR spectroscopy resulted in the unambiguous conclusion of HA secondary structures and was described in details by, among others, Welti,[33] Scott,[34] Heatley,[35] and Cowman.[36] Some of those examples have been referenced and discussed in Chapter 1, regarding their significance in understanding the secondary structure of HA.

One of the challenges in the study of chemical reactions and modifications is to gather detailed, useful information to interpret the exact structure of the reaction product. The confirmation of the anticipated product structure is all the more important for the reactions involving HA, as its chemical behavior is often complicated and unpredictable. Kuo et al.[37,38] studied the chemical reaction of HA, primary amine, and carbodiimide. The carbodiimide was expected to form an unstable intermediate O-acylurea of HA, which would further react with the primary amine to form the final product HA amide. The proton NMR data, however, indicated otherwise. The chemical shift of protons corresponding to the methylene protons next to the amino group (~2.6 ppm) was not detected, excluding the possibility of the forming of an HA amide. Instead, a strong singlet at 2.18 ppm and other signals indicated that the carbodiimide was covalently bonded to HA in forming HA-acylurea. NMR provided key evidence to the mechanism of carbodiimide–HA reactions, and opened an avenue into a series of new HA modification products through reactions of HA with designed carbodiimides.

The HA acylurea structure was further confirmed by Pouyani et al.[39] by solid-state NMR. The ^{13}C- and ^{15}N-labeled carbodiimides were synthesized to react with HA. Cross polarization combined with magic angle spinning (CP-MAS) gives well-resolved ^{13}C or ^{15}N-NMR spectra, with nearly quantitative signals. The use of CP-MAS NMR spectroscopy extended to the study of other chemical modifications of HA, as in a study of Passerini and Ugi reactions by de Nooy et al.[40]

In spite of the recognition of its contribution to HA structural elucidation, NMR has limited use as a routine assay in the manufacture of HA products. This has to do with the cost associated with the equipment and its relatively low sensitivity. Practically no single analytical method can be found to be sufficient all the time when used alone. A combination of carefully selected methods, such as mentioned in this chapter, with or without NMR can often satisfy the need for HA analysis under various conditions.

4.2.4 Mass Spectrometry

Mass spectrometry (MS) is a powerful and sensitive method. A mass spectrometer converts components of a sample into rapidly moving gaseous ions and resolves them on the basis of their mass-to-charge ratios. To identify the exact size of HA oligosaccharide fractions from hyaluronidase degradation, various kinds of MS methods have been experimented and reported. Examples are the publications by Price et al. (electrospray-ionization MS),[41] by Yeung et al. (matrix-assisted laser desorption ionization MS),[42] and by Mahoney et al. (a combination of the above two methods).[43] Mass spectrometry often follows a separation process such as gas liquid chromatography (GLC), high performance liquid chromatography (HPLC), or electrophoresis, as the purity of the sample is critical for MS analysis. Tawada et al.[44] used electrospray-ionization MS to analyze large-scale (200 g HA) preparation of HA oligosaccharides ranging from 4-mers to 52-mers.

4.2.5 Light Scattering

Light scattering (LS) is one of the absolute methods for the characterization of HA MW. The other absolute methods are osmometry[45] and sedimentation.[46,47]

Absolute methods give direct information of polymer MW without the need for calibration by other standards, whereas nonabsolute methods such as chromatography and viscometry rely on absolute methods as points of reference. For instance, the Mark equation (also named the Mark-Howink equation) widely used in viscometry was calibrated against the light-scattering measures, as described in Chapter 3.[48,49] In light-scattering analysis, a single-frequency polarized laser beam is used to illuminate a solution containing the macromolecules of interest. The measured intensity carries information of the molar mass, whereas the angular dependence carries information related to the size of the macromolecule.

Laurent et al. first reported in 1955 the use of LS in the study of HA[50] (wavelength 4358 Å, angular distribution of scattered light 35–135°). The study concluded that HA "behaves like a somewhat rigid coil" and that the MW of the HA from human umbilical cord was 2.8 to 4.3 million.

The application of SEC/MALLS (i.e., size-exclusion chromatography coupled online to a multiangle laser light-scattering photometer) is becoming increasingly popular and has been used to analyze some commercially available high-MW HA samples.[51] SEC/MALLS has the benefit of combining good separation by SEC and precise measure of MW by MALLS. The MALLS detectors simultaneously measure the intensity of light scattered by polymers in solutions at different angles. Typically, such a detector has a lower detection limit for molar mass in the range of a few hundred dalton and a limit for a radius of gyration of 10 nm.

Hokputsa et al.[47] compared the results of MW of degraded HA (by ascorbic acid and hydrogen peroxide) using SEC/MALLS and sedimentation equilibrium. With the SEC/MALLS method, HA is measured at very low concentration on the order of 10^{-5} g/ml, thus avoiding the problem with "solution nonideality" arising from charge effects and excluded volume effects that increase with concentration. In the case of sedimentation equilibrium, it is necessary to perform measurement at several concentrations and extrapolate to zero concentration. Therefore, SEC/MALLS is faster and more efficient than the sedimentation equilibrium method.

The LS technique is noninvasive to the samples being tested and can avoid errors resulting from the shearing force to which HA would be subjected during viscosity measurement. However, the operation of LS may be prone to error in its own ways. The refractive index increment (dn/dc) is an essential parameter in determining MW and is sensitive to impurities and errors in measuring HA concentrations. Because the square of dn/dc is used in the calculation of light-scattering equations to determine the MW, the error in refractive index increment is magnified. The dn/dc values of HA as reported in previous literature ranged from 0.140 to 0.183 ml/g.[46] In the study by Hokputsa et al.,[47] the dn/dc of HA was experimentally determined as 0.167 ml/g (reported by Hokputsa as 0.167 g/ml; as n is unitless, the unit of n/c is the inversion of the unit of c, and therefore not g/ml, but ml/g).

4.2.6   Liquid Chromatography

4.2.6.1   Ion-Exchange Chromatography

In cation-exchange chromatography, sodium ion of NaHA was replaced by other cations such as proton[52] and quaternary ammonium ions.[53,54] Anion-exchange chromatography was used to purify HA from associated proteins.[55,56] For measuring the MW distribution of HA, the difficulty in gaining good separation with ion-exchange chromatography increases as the MW of HA increases.

Nebinger[57] studied the methods of anion-exchange chromatography for the separation HA oligosaccharides as degradation products of various hyaluronidases and found that DEAE Sephacel (formate form) could separate a range of HA oligosaccharides up to decasaccharide with good resolution and high yield. For odd-numbered HA oligosaccharides, the ion-exchange method has an obvious advantage over size-exclusion method, because, for example, (GlcNAc-GlcUA-GlcNAc) and (GlcUA-GlcNAc-GlcUA) hardly differ in MW, but the latter has twice the charge as the former.

By using high-performance anion-exchange chromatography with a pulsed amperometric detector, Zhang et al.[58] were able to determine the molecular mass of an anionic polysaccharide, such as sialic acid polymers, up to a degree of polymerization of approximately 80. Karlsson recently reported[59] a method using "strong anion exchange column with quaternary amine functional group (PL-SAX-4000) and a mobile phase consisted of sodium phosphate and/or sodium sulfate." The method has the capacity to determine the molecular mass distribution for HA in the range of 0.1–5 million Da. That is equivalent to a degree of polymerization of approximately 250–12,500 disaccharide units. The seven HA samples used in the study were from Pharmacia, and their MW measure was calibrated against a low-angle laser light-scattering photometer.

4.2.6.2   Size Exclusion Chromatography

Size exclusion chromatography (SEC), also called gel-filtration and gel-permeation chromatography, separates HA molecules according to their size. The size of the molecules is related to the different retention time of molecules by the pores of column material. The larger molecules are eluted before the smaller molecules, opposite to what is usually observed for ion-exchange chromatography.[60]

SEC is a convenient and effective method to characterize the molecular size of HA, provided the MW is not too high. Miyazaki used ultrasonication to reduce HA MW from over a million to the range of one to a few hundred thousand dalton. The fractions of HA were then separated by three TSK columns.[61]

In general, the power of SEC to resolve the small differences of the size of high-MW HA ($>10^6$) is quite limited. It often lacks the reproducibility from lab to lab, probably because of its sensitivity to the exact condition of use, such as shear force and sample concentration, analogous to the situation HA may encounter in viscometry. Successful application of SEC to determine the MW of HA at or above 10^6 Da has been rare. However, a few examples are noted. Terbojevich et al. reported the use of a Shodex OHPAK B 806 column to evaluate the MW distribution of HA up to 1 million dalton.[62] Beaty et al.[107] reported the measure of HA MW up to 2 million Da, using a TSK G6000 PW column. The concentrations of HA samples were between 0.1 and 0.3 mg/ml.

4.2.7 Electrophoresis

Agarose gel electrophoresis,[63] cellulose acetate electrophoresis,[64] and capillary electrophoresis[65,66] have been developed for the separation and quantifying of HA of a wide range of MW distribution. Electrophoresis is the movement of molecules through a gel or a fluid under the influence of an electric field. The separation of HA molecules from other molecules or among polydispersed HA molecules is based on the principle that electrophoretic mobility of a molecule is proportional to its electrical charge and inversely proportional to its molecular mass.

Lee et al.[63] established a linear relationship between electrophoretic mobility and the logarithm of the weight-average MW of HA over the range of 0.2–6.0 million Da. The pattern of HA separation can be visualized by staining with Alcian blue after HA is electro-transferred from the agarose gel to a nylon membrane. The MW of two liquid connective tissues was measured. The MW of normal human knee joint synovial fluid was 6–7 million Da, and 5–6 million Da for the owl monkey vitreous.

Nakano et al.[64] used the cellulose acetate electrophoretic method to monitor the HA extraction from combs and wattle tissues. After limited steps of purification such as acetone drying and proteolysis, the crude HA can be assayed without deproteinization. This effectively reduced the time to measure HA content to less than a day.

Platzer et al.[65] used high-performance capillary electrophoresis to quantify HA in pharmaceutical formulations. The method involves a fused silica capillary with bubble detection cell and an on-column diode array detector (190–600 nm). Linearity was established between 0.01 and 5.0 mg/ml. The lower limit of detection is 10 μg/ml, as compared to 7.5 μg/ml for carbazole assay. The relative standard deviation of the capillary electrophoresis method (9.8%) was much better than for the carbazole method (18.4%).

The above examples illustrated the utility of electrophoresis. The method is quite robust, and the sample normally does not require extensive purification before testing. HA as a natural component of connective tissues, or as an ingredient in pharmaceutical formulations, can be separated, confirmed, and quantified in a single procedure.

4.2.8 Osmometry

Osmotic pressure can develop when two solutions are separated by a semipermeable membrane that allows only the solvent and smaller solute molecules to pass through. The osmolality (mOsm/kg) of an HA aqueous solution is equivalent to its osmolarity (mOsm/liter), because the water density is one. In ISO-OVD, the range of osmolality for ophthalmic used was defined (200–400 mOsm/kg), because osmolality is directly related to the biocompatibility and safety of the HA products. For example, cells may rupture (red blood cells) when placed in hypotonic medium (water) and shrink in hypertonic medium (salt). The methods to measure osmolarity are described in USP. The freezing-point depression method is recommended, and the vapor pressure method is less frequently employed. The contribution of a solute to the osmolarity of a solution, like vapor pressure and freezing point, is colligative in nature. It depends mainly on the number of molecules in the solution. When measured at a certain concentration, MW information may be derived from the osmolarity data.

The osmolarity of many HA products is largely attributed to the buffer salt in the formulation. The osmolarity of 0.9% NaCl, according to *USP <785> Osmolarity*, is about 286 mOsm per liter. There are usually larger numbers of Na^+ and Cl^- ions in the solution from the dissociation of NaCl salt than the Na^+ and HA ions from NaHA. However, the contribution of HA to osmolarity should not be underestimated. The Donnan effect is a significant factor for polyelectrolytes, such as NaHA. Donnan equilibrium is asymmetric concentration distribution of the small ions at, for instance, the two sides of the semipermeable membrane for dialysis. When macromolecules are negatively charged, they hold many small counter cations, such as Na^+ to their side of the membrane, thus creating an osmotic pressure that remains after the dialysis.

The osmotic pressure caused by a colloidal system (Donnan pressure) such as an HA solution is sensitive to the change of polymer concentration. Bothner[67] stated that "HA solution exhibits a highly non-ideal colloid osmotic pressure behavior. That is, the colloid osmotic pressure increases very rapidly in increasing concentration." Therefore, the presence of HA in the extracellular matrix is largely responsible for the water balance and hydration of connective tissues.

Dick et al.[4] carried out a study comparing the osmolality of over 20 viscoelastic products (OVD) including HA and HPMC. The freezing point depression method was used and the range of osmolality of the OVDs was found between 279 (Biolon) and 376 (HYA-Ophtal) mOsm/kg. Other examples were Amvisc Plus (335), AMO Vitrax (284), Healon (295), Healon GV (312), Healon 5 (322), and Viscoat (340).

4.3   Natural Polymeric Impurities

Proteins, nucleic acid, and sulfated glycosaminoglycans are often found in HA tissue extracts, even after multistep purification processes. A pure HA formulation shows a baseline in the ultraviolet region above 230 nm. The absorbance at 260 nm can be used as a measure of the presence of nucleic acid, as recommended by EP. Protein solutions show a strong ultraviolet light absorption at 280 nm caused by the aromatic rings of tyrosine and tryptophan.[68]

For total protein content, EP specifies an upper limit of 0.3% of the HA dry weight. Traditional Lowry assay[68,69] can serve the purpose. The Lowry assay is sensitive because it uses both the biuret reaction (Cu^{++} and peptide) and the reaction between tyrosine (a phenolic component of protein) and Folin reagent (a complex of phosphotungstomolybdic acid). Folin reagent is reduced from its original golden yellow color to deep blue. The absorption peak was observed at approximately 750 nm. The application of the Lowry method to compare the protein content of HA-based products has been reported.[70]

Swann[71] conducted a study of the acid hydrolysis of the HA protein complex from rooster combs. The protein components associated with HA were measured by ion exchange chromatography. Five amino acids — serine, glycine, alanine, aspartic acid, and glutamic acid — accounted for 65% of the total amino acid present. The total protein content was 0.35% of the HA weight.

The total sulfated glycosaminoglycans can be measured using the EP method (perchloric acid/$BaCl_2$). Because galactosamine is a component of some sulfated glycosaminoglycans, such as chondroitin sulfate, analysis of galactosamine content by GLC method[72] mentioned in Chapter 2 can also provide information on the purity of HA versus other glycosaminoglycans.

4.4   Other Bioassays

4.4.1   Eye Tolerance Assays

4.4.1.1   Owl Monkey Eye Model

Viscoelastic formulations containing hyaluronic acid and collagen were injected into monkey vitreous to examine the feasibility of using such formulations as vitreous replacement. The eyes of South American monkeys, such as owl monkeys (*Aotus trivirgatus*), were chosen because their vitreous is fluid and can be aspirated from the eye through a number 26 needle.[73–75]

The intravitreous HA injection into monkey eyes was established as a primate model for the testing of the ocular tolerance of Healon and other HA products. In those studies, about half of the vitreous (vitreous is a noun, vitreous is an adjective) was replaced by various HA products. A maximum response of the eye by an increased number of leukocytes in the aqueous and vitreous was noted at 48 hours.[76,77]

The results indicated that the owl monkey eye (OME) assay was quite sensitive in detecting the existence of the inflammatory substances. The positive controls and the inflammatory batches of HA showed high white blood cell counts ($\geq$1000 in 1 mm^3 vitreous), whereas the noninflammatory batches of HA varied between 0 and 200 cells/mm^3 (20 ± 3). Therefore, the difference of average white blood cell (WBC) counts between inflammatory and noninflammatory is more than an order of magnitude.

There are two major factors contributing to the inflammatory response. The first is the trauma by the surgery itself, which according to Balazs, can be accountable for WBC counts of 0–70. The second factor is the amount of inflammatory substance injected into the vitreous body. It is noteworthy that inflammatory responses appear to have a threshold below which there is no inflammation, and above which cellular response can be amplified to a greater magnitude. The threshold of WBC counts for inflammatory response was determined, more or less arbitrarily, to be at 200 per cubic millimeter of aqueous or vitreous humor.[78] In addition to the WBC counts, the overall tissue reaction is included in the criteria as well. Balazs implanted 1 ml of a 1% solution of NaHA (in physiological buffer) in the vitreous to replace approximately one-half the existing vitreous body. There was no significant cellular infiltration of the vitreous and anterior chamber, no flare in the aqueous humor, no haze or flare in the vitreous, and no pathological changes to the cornea, lens, iris, retina, and choroid of the owl monkey eye. HA batches passing such tests were considered noninflammatory.

It was further concluded that the assay was practical because the "animal could be used repeatedly after given sufficient time to recover." This was based on the fact that the inflammatory reaction was not more severe or longer lasting when the implantation was repeated on the same eye up to six times in 3 to 7-month intervals.[76] The OME assay became the routine tests for the manufacture of some of the early brands of HA viscoelastics for many years.

Note the obvious drawback of using methods that involve primates: it is getting more and more difficult to justify the continued use of OME tests for the reasons of compassion as well as cost. In the OME assay, there is no linear relationship between the amount of material and the white blood cell counts. Furthermore, it is debatable whether the leukocyte counts in the eye after the HA injection are truly indicative of the reactions other tissues and organs may have to the material.

4.4.1.2   Rabbit Eye Model

In the early development stages of Healon, Amvisc, and Viscoat, rabbit eye tests were already experimented with and reported on. For obvious reasons, rabbits have been increasingly used as substitutes for monkeys to test the tolerance of HA material.[79–83]

Unlike owl monkey, the vitreous of rabbits is not quite liquid. In a typical procedure described by Skelnik and Lindstrom,[84] a 4–5-mm incision is made through the conjunctiva, and a 19-gauge needle is needed to remove the vitreous from New Zealand white rabbits.

4.4.2   Limulus Amebocyte Lysate Assay

The Limulus Amebocyte Lysate (LAL) assay is an *in vitro* assay for endotoxins — lipopolysaccharides originating from Gram-negative bacteria. The lipopolysaccharides are components of the outer cell wall and are released on autolytic disintegration of bacteria.

There are two major types of LAL test — gel-clot technique and photometric technique (turbidimetric method and chromogenic method). The gel-clot technique is based on the gel formation as the result of reaction of endotoxins with Limulus Amebocyte Lysate, which is obtained from the aqueous extracts of circulating amebocytes (blood corpuscles) of the horseshoe crab (*Limulus polyphemus*). "In case of dispute, the final decision is based on the gel-clot techniques," according to USP <85>.

Endotoxin limit is determined by the type of tissue with which the "device" is in contact. USP has set the endotoxin limit of medical devices to no more than 20.0 endotoxin units (EU) per device "except that for those medical devices in contact with the cerebrospinal fluid the limit is not more than 2.15 USP endotoxin units per device" (USP <161>, *Transfusion and Infusion Assemblies and Similar Medical Devices*). Because HA is not a typical device, the endotoxin limit for HA was adjusted to reflect its relevant characteristics such as volume or weight. For OVD, a lower limit of detection of 0.5 EU/ml is mandatory. "Any product that exceeds a bacterial endotoxin limit of 0.5 endotoxin unit (EU) per milliliter fails the test" (6.2.3. Bacterial endotoxin test in ISO/15798).[2] Similarly, the endotoxin limit for ocular use as required by EP is 0.05 IU/mg or 0.5 IU/10 mg (one USP-EU is equal to one international unit of endotoxin, <85>, *Bacterial Endotoxin Test*). The EP and ISO limits for endotoxin of HA for ophthalmic use are basically consistent. Take a typical ophthalmic HA formulation, 10 mg/ml HA, as an example. Theoretically, if the endotoxin level of solid HA falls below 0.05 IU/mg, then the endotoxin level of a formulation containing 10 mg of HA should be less than 0.5 EU units/ml.

Dick et al. tested 25 commercially available OVDs using the EP/USP clotting method. In that study, the lower limit of detection was 1.2 EU/ml, probably because of high viscosity of some of the OVDs and therefore high dilution factors. The results showed that many OVDs had rather high counts of EU. It was found that 16 OVDs had an endotoxin content under 1.2 EU/ml, 5 OVDs had 1.2–24 EU/ml, and 4 had more than 24 EU/ml.[5]

The relation between adverse events of OVDs and higher endotoxin levels has been established. For example, in 1997, the use of Microvisc (in Sweden) was found to cause endophthalmitis-like reactions that occurred after complication-free cataract surgeries. Tests of OVD samples by the Swedish authorities in many cases yielded an endotoxin content of up to 100 EU/ml.[5,85]

4.5 Lipoteichoic Acid

Lipoteichoic acid (LTA) is a phosphate-containing polymer, isolated from cell walls and membranes of Gram-positive bacteria.[86] LTA is associated with Gram-positive bacteria, whereas lipopolysaccharides is associated with Gram-negative bacteria.

It is required by European Pharmacopoeia[1] that when HA is produced "by fermentation of Gram positive bacteria, the process must be shown to reduce or eliminate pyrogenic or inflammatory components of the cell wall." This may be to some extent applicable to HA extraction from animal tissues such as rooster combs as a result of bacterial contamination of the tissue. In spite of the importance of controlling the level of LTA, the standard for LTA limit in HA product has not been established, nor is there a widely accepted method to test it. However, assays based on LTA antibodies have been reported. Hogg and Manning developed an enzyme-linked immune sorbent assay to measure bacterial surface LTA.[87] Tadler et al. developed a sandwich immunoassay for the detection of LTA from whole blood.[88]

4.6 Chemical Methods of Radiolabeling

Radiolabeled HA has been used mainly for the purpose of studying the metabolism of HA.[89–95] The advantage of using radiolabeled HA in those studies is obvious. Radiolabeled exogenous HA can be easily distinguished from the endogenous HA of the subject of the study. Therefore, it is not necessary to compare data of exogenous HA with an ever-changing baseline of endogenous HA. For the study of absorption, distribution, and elimination of exogenous HA, using radiolabeled HA has been the method of choice.

There are two major methodologies for radiolabeling HA, one being chemical and the other biosynthetic. Chemical methods involve chemical reactions of HA with radioactive reagents containing ^{3}H, ^{14}C, ^{125}I, or ^{11}C. Sometimes it requires that the HA be chemically modified before a radioactive moiety can be introduced. Biosyntheses use cell culture or fermentation processes with radioactive precursors (^{3}H or ^{14}C) such as glucose, acetate, and so on.[96,97] Biosynthetic techniques can afford radioactive HA structurally identical to the natural HA. However, the processes are laborious and do not have much control of the MW of the HA.

The following is a summary of the several chemical methods that have been investigated to radiolabel HA.[98–105] Each method has its limitations as well as values for certain intended applications.

4.6.1 Tritium

4.6.1.1 *Exchange Reaction*

Tritium gas exchange reaction has limited use in the tritium labeling of HA. It requires long exposure of HA solution to tritium and may cause serious degradation of HA.[98] Moreover, much of the tritium labeled to HA by exchange reaction is associated with hydroxyl groups of HA, and therefore labile. This method is applicable mainly to the making of radioactive HA oligosaccharides.[99]

4.6.1.2 *Acetylation*

The methyl tritium in the acetyl group is not labile. The acetylation of HA by [^{3}H] acetic anhydride requires that HA first be deacetylated by reaction with hydrazine.[100] The effect of such treatment on the chain size of HA (from rooster comb) was not investigated in detail, except that the labeled HA was excluded from Sephadex G-200. The labeled HA appeared to maintain certain biological properties of native HA, with regard to the polyanion property, interaction with proteoglycan. The specific activity of the [^{3}H] acetyl labeled HA was $0.3–0.42 \times 10^6$ cpm/µg uronic acid.[100]

4.6.1.3 *Reduction of Aldehyde*

The reduction of the aldehyde functions by [^{3}H]NaBH$_4$ can introduce stable tritium label. The aldehyde function of HA does not occur in significant quantities, especially for high-MW HA, because it only exists at the reducing end, in a form of equilibrium between the aldehyde form (open ring) and the hemiacetal form (closed ring). Orlando's tritiation method[105] is characterized by the partial oxidation of HA to generate aldehyde functional groups on HA molecule, followed by reduction with [^{3}H]NaBH$_4$. However, the biological ramification of such a ring opening by oxidation of HA is not exactly clear. For low-MW HA, the partial oxidation method may not be necessary, as the degree of labeling is dependent on the number of end groups. The advantage of labeling only at the reducing end is to keep the perturbation of HA structure to the minimum.

4.6.2 Carbon 14

The reaction of HA reducing end groups with [^{14}C] potassium cyanide has been reported by Swann et al.[101] The method was developed to measure HA MW, which is inversely related to the number of the labeling sites available at the reducing end of HA. It was not intended for applications in animal studies, because of the toxic properties of the radiolabeled HA.

4.6.3   Iodine 125

Radioiodination involves at least two steps of HA modification, first to introduce a functional group reactive to iodine or its salt, followed by iodination. For example, tyramine was first introduced to the reducing end of HA, and the formed Schiff base was reduced with $NaCNBH_3$, followed by $Na^{125}I$ treatment.[102] Another example[103] was the reaction of tyrosine with CNBr-activated HA, followed by treatment with ^{125}I. These methods require purification at each step with low yields.

The iodination generally can provide a labeled-HA final product with high specific radioactivity. However, iodine is not a natural element of an HA molecule, and one should be aware of the structural change of HA caused by iodination and its potential effect on the biological functions of HA. A rat study by Gustafson et al.[103] showed that although the binding of iodinated HA to liver endothelial cells was the same as biosynthetically labeled HA, the wash out of radioactivity from the cells for iodinated HA was much slower than that of the biosynthetically labeled tritium HA.

4.6.4   Carbon 11

^{11}C labeled HA can be synthesized by the reaction of HA with $[^{11}C]$cyanogen bromide (CNBr), as reported by Westerberg et al.[104] The chemistry of HA with CNBr has been described in Chapter 2. The radioactive precursor of $[^{11}C]$cyanogen was $[^{11}C]$carbon, which was produced by a cyclotron. The synthesis time of radiolabeled HA was 8–10 minutes, counted from the end of the bombardment. The radiolabeling of HA must be conducted at the application sites because of the very short half-life of ^{11}C (20.3 minutes).

Linqvist et al. were the first to use ^{11}C labeled HA to study the liver uptake rate of HA by patients with liver dysfunction, which was found to be significantly lower than the healthy patients. The rate of the uptake of ^{11}C labeled HA was determined by positron emission tomography.[106]

References

1. Sodium Hyaluronate Natrii hyalurona, in *European Pharmacopoeia Supplement*, pp 1190–1193, European Directorate for the Quality of Medicines of the Council of Europe, Strasbourg, 2000.
2. Ophthalmic Implants — Ophthalmic Viscosurgical Devices, *ISO 15798* (2001).
3. Arshinoff, S. A., New terminology: ophthalmic viscosurgical devices, *J. Cataract Refract. Surg.*, 26, 627–628 (2000).
4. Dick, H. B., Augustin, A. J., and Pfeiffer, N., Osmolality of various viscoelastic substances: comparative study, *J. Cataract Refract. Surg.*, 26, 1242–6 (2000).
5. Dick, H. B., Augustin, A. J., Pakula, T., and Pfeiffer, N., Endotoxins in ophthalmic viscosurgical devices, *Eur. J. Ophthalmol.*, 13, 176–184 (2003).
6. Torngren, L., Lundgren, B., and Madsen, K., Intraocular pressure development in the rabbit eye after aqueous exchange with ophthalmic viscosurgical devices, *J. Cataract Refract. Surg.*, 26, 1247–1252 (2000).

7. Magnusson, B. and Kligman, A. M., The identification of contact allergens by animal assay. The guinea pig maximization test, *J. Invest. Dermatol.,* 52, 268–276 (1969).

8. Ames, B. N., McCann, J., and Yamasaki, E., Methods for detecting carcinogens and mutagens with the Salmonella/mammalian-microsome mutagenicity test, *Mutat. Res.,* 31, 347–364 (1975).

9. Galloway, S. M., Bloom, A. D., Resnick, M., Margolin, B. H., Nakamura, F., Archer, P., and Zeiger, E., Development of a standard protocol for *in vitro* cytogenetic testing with Chinese hamster ovary cells: comparison of results for 22 compounds in two laboratories, *Environ. Mutagen,* 7, 1–51 (1985).

10. Orr, S. F. D., Infra-red spectroscopic studies of some polysaccharides, *Biochim. Biophys. Acta,* 14, 173–181 (1954).

11. Cael, J. J., Isaac, D. H., Blackwell, J., and Koenig, J. L., Polarized infrared spectra of crystalline glycosaminoglycans, *Carbohydr. Res.,* 50, 169–179 (1976).

12. Atkins, E. D., Phelps, C. F., and Sheehan, J. K., The conformation of the mucopolysaccharides, hyaluronates, *Biochem. J.,* 128, 1255–1263 (1972).

13. Guss, J. M., Hukins, D. W., Smith, P. J., Winter, W. T., and Arnott, S., Hyaluronic acid: molecular conformations and interactions in two sodium salts, *J. Mol. Biol.,* 95, 359–84 (1975).

14. Mitra, A. K., Raghunathan, S., Sheehan, J. K., and Arnott, S., Hyaluronic acid: molecular conformations and interactions in the orthorhombic and tetragonal forms containing sinuous chains, *J. Mol. Biol.,* 169, 829–59 (1983).

15. Gilli, R., Kacurakova, M., Mathlouthi, M., Navarini, L., and Paoletti, S., FTIR studies of sodium hyaluronate and its oligomers in the amorphous solid phase and in aqueous solution, *Carbohydr. Res.,* 263, 315–326 (1994).

16. Bitter, T. and Muir, H. M., A modified uronic acid carbazole reaction, *Anal. Biochem.,* 4, 330–334 (1962).

17. Gold, E. W., A simple spectrphotometric method for estimating glycosaminoglycan concentrations, *Anal. Biochem.,* 99, 183–188 (1979).

18. Morgan, W. T. J. and Elson, L. A., A colorimetric method for the determination of N-acetylglucosamine and N-acetylchondrosamine, *Biochem. J.,* 28, 988–995 (1934).

19. Aminoff, D., Morgan, W. T. J., and Watkins, W. M., The action of dilute alkali on the N-acetylhexosamines and the specific blood-group mucoids, *Biochem J.,* 51, 379–389 (1952).

20. Reissig, J. L., Strominger, J. L., and Leloir, L. F., A Modified colorimetric methods for the estimation of N-acetylamino sugars, *J. Biol. Chem.,* 217, 959–966 (1955).

21. Asteriou, T., Deschrevel, B., Delpech, B., Bertrand, P., Bultelle, F., Merai, C., and Vincent, J. C., An improved assay for the N-acetyl-D-glucosamine reducing ends of polysaccharides in the presence of proteins, *Anal. Biochem.,* 293, 53–59 (2001).

22. Jourdian, G. W., Wolfman, M., Sarber, R., and Distier, J., A specific, sensitive method for the determination of hyaluronate, *Anal. Biochem.,* 96, 474–480 (1979).

23. Hascall, V. C., Riolo, R. L., Hayward Jr. J., and Reynolds, C. C., Treatment of bovine nasal cartilage proteoglycan with chondroitinases from *Flavobacterium heparinum* and *Proteus vulgaris, J. Biol. Chem.,* 247, 4521–4528 (1972).

24. Adams, M. E. and Muir, H., An automated version of the periodate-thiobarbituric acid assay for analysis of Δ-4,5 unsaturated uronic acids and its application to the assay of hyaluronic acids and chondroitin sulfates, *Anal. Biochem.,* 109, 426–431 (1980).

25. Li, X. Q., Thonar, E. J. M. A., and Knudson, W., Accumulation of hyaluronate in human lung carcinomam as measured by a new hyaluronate ELISA, *Connect Tissue Res.*, 19, 243–254 (1989).

26. Chichibu, K., Matsuura, T., Shichijo, S., and Yokoyama, M. M., Assay of serum hyaluroic acid in clinical application, *Clin. Chim. Acta*, 181, 317–324 (1989).

27. Delpech, B., Bertrand, P., and Maingonnat, C., Immunoenzymoassay of the hyaluronic acid-hyaluronectin interaction: application to the detection of hyaluronic acid in serum of normal subjects and cancer patients, *Anal. Biochem.*, 149, 555–565 (1985).

28. Goldberg, R. L., Enzyme-linked immunosorbent assay for hyaluronate using cartilage proteoglycan and an antibody to keratan sulfate, *Anal. Biochem.*, 174, 448–58 (1988).

29. Lindqvist, U., Chichibu, K., Delpech, B., Goldberg, R. L., Knudson, W., Poole, A. R., and Laurent, T.C., Seven different assays of hyaluronan compared for clinical utility, *Clin. Chem.*, 38, 127–132 (1992).

30. Chichibu, K., Method of assaying high molecular hyaluronic acid and kit of reagents for such assay, U.S. Patent 5,019, 498, (1991).

31. *Hyaluronic Acid (HA) Quantitative Test Kit, for Investigational Use Only*, Corgenix, Inc., Westminster, Co. http://www.corgenix.com, (2001).

32. Takei, Y. G., Honma, T., and Ito, A., Quantitation of hyaluronic acid in serum with automated microparticle photometric agglutination assay, *J. Immunoassay Immunochem.*, 23, 85–94 (2002).

33. Welti, D., Rees, D. A., and Welsh, E. J., Solution conformation of glycosaminoglycans: assignment of the 300-MHz 1H-magnetic resonance spectra of chondroitin 4-sulphate, chondroitin 6-sulphate, and hyaluronate, and investigation of an alkali-induced conformation change, *Eur. J. Biochem.*, 94, 505–514 (1979).

34. Scott, J. E., Heatley, F., and Hull, W. E., Secondary structure of hyaluronate in solution. A 1H-N.M.R. investigation at 300 and 500 MHz in [2H6]dimethyl sulphoxide solution, *Biochem. J.*, 220, 197–205 (1984).

35. Heatley, F. and Scott, J. E., A water molecule participates in the secondary structure of hyaluronan, *Biochem. J.*, 254, 489–493 (1988).

36. Cowman, M. K., Cozart, D., Nakanishi, K., and Balazs, E. A., ^{1}H NMR of glycosaminoglycans and hyaluronic acid oligosaccharides in aqueous solution: the amide proton environment, *Arch. Biochem. Biophys.*, 230, 203–212 (1984).

37. Kuo, J. W., Swann, D. A., and Prestwich, G. D., Chemical modification of hyaluronic acid by carbodiimides, *Bioconjug. Chem.*, 2, 232–241 (1991).

38. Kuo, J. W., in *Synthesis and Properties of Hyaluronic Acid Modified by Designed Carbodiimides, Department of Chemistry*, State University of New York, Stony Brook, 1989, pp. 1–131.

39. Pouyani, T., Kuo, J.-W., Harbison, G. S., and Prestwich, G. D., Solid-state NMR of N-acylureas derived from the reaction of hyaluronic acid with isotopically-labeled carbodiimides, *J. Am. Chem. Soc.*, 114, 5972–5976 (1992).

40. de Nooy, A. E., Capitani, D., Masci, G., and Crescenzi, V., Ionic polysaccharide hydrogels via the Passerini and Ugi multicomponent condensations: synthesis, behavior and solid-state NMR characterization, *Biomacromolecules*, 1, 259–267 (2000).

41. Price, K. N., Tuinman, A., Baker, D. C., Chisena, C., and Cysyk, R. L., Isolation and characterization by electrospray-ionization mass spectrometry and high-performance anion-exchange chromatography of oligosaccharides derived from hyaluronic acid by hyaluronate lyase digestion: observation of some

heretofore unobserved oligosaccharides that contain an odd number of units, *Carbohydr. Res.*, 303, 303–311 (1997).

42. Yeung, B. and Marecak, D., Molecular weight determination of hyaluronic acid by gel filtration chromatography coupled to matrix-assisted laser desorption ionization mass spectrometry, *J. Chromatography*, 852, 573–581 (1999).

43. Mahoney, D. L., Aplin, R. T., Calabro, A., Hascall, V. C., and Day, A. J., Novel methods for the preparation and characterization of hyaluronan oligosaccharides of defined length, *Glycobiology*, 11, 1025–1033 (2001).

44. Tawada, A., Masa, T., Oonuki, Y., Watanabe, A., Matsuzaki, Y., and Asari, A., Large-scale preparation, purification, and characterization of hyaluronan oligosaccharides from 4-mers to 52-mers, *Glycobiology*, 12, 421–426 (2002).

45. Christiansen, J. A. and Jensen, C. E., An inverted microosmometer and its use for the determination of molecular weight of some specimens of potassium hyaluronate, *Acta Chem. Scand.*, 7, 1247–54 (1953).

46. Lapcik, L., Jr., Lapcik, L., De Smedt, S., Demeester, J., and Chabrecek, P., Hyaluronan: preparation, structure, properties, and applications, *Chem. Rev.*, 98, 2663–2684 (1998).

47. Hokputsa, S., Jumel, K., Alexander, C., and Harding, S. E., A comparison of molecular mass determination of hyaluronic acid using SEC/MALLS and sedimentation equilibrium, *Eur. Biophys. J.*, 32, 450–456 (2003).

48. Laurent, T. C., Ryan, M., and Pletruszkiewicz, A., Fractionation of hyaluronic acid, the polydispersity of hyaluronic acid from the bovine vitreous body, *Biochim. Biophys. Acta.*, 42, 476–485 (1960).

49. Ueno, Y., Tanaka, Y., Horie, K., and Tokuyasu, K., Low-angle laser light scattering measurements on highly purified sodium hyaluronate from rooster comb, *Chem. Pharm. Bull.*, 12, 4971–4975 (1988).

50. Laurent, T. C. and Gergely, J., Light scattering studies on hyaluronic acid, *J. Biol. Chem.*, 212, 325–333 (1955).

51. Soltes, L., Mendichi, R., Lath, D., Mach, M., and Bakos, D., Molecular characteristics of some commercial high-molecular-weight hyaluronans, *Biomed. Chromatogr.*, 16, 459–462 (2002).

52. Laurent, T. C., Studies on hyaluronic acid in the vitreous body, *J. Biol. Chem.*, 216, 263–271 (1955).

53. Della Valle, F. and Romeo, A., Esters of hyaluronic acid, U.S. Patent 4,851,521, (1989).

54. Luo, Y. and Prestwich, G. D., Hyaluronic acid-N-hydroxysuccinimide: a useful intermediate for bioconjugation, *Bioconjug. Chem.*, 12, 1085–1088 (2001).

55. Berman, E. R., Isolation of bovine vitreous hyaluronic acid on diethylaminoethyl-sephadex, *Biochim. Biophys. Acta*, 58, 120–122 (1962).

56. Swann, D. A., Studies on hyaluronic acid. I. The preparation and properties of rooster comb hyaluronic acid, *Biochim. Biophys. Acta*, 156, 17–30 (1968).

57. Nebinger, P., Comparison of gel permeation and ion-exchange chromatographic procedures for the separation of hyaluronate oligosaccharides, *J. Chromatography*, 320, 351–359 (1985).

58. Zhang, Y., Inoue, Y., Inoue, S., and Lee, Y. C., Separation of oligo/polymers of 5-N-acetylneuraminic acid, 5-N-glycolylneuraminic acid, and 2-Keto-3-deoxy-glycero-galacto-nononic acid by high-performance anion-exchange chromatography with pulsed amperometric detector, *Anal. Biochem.*, 250, 245–251 (1997).

59. Karlsson, G. and Bergman, R., Determination of the distribution of molecular masses of hyaluronate by high-performance anion-exchange chromatography, *J. Chromatography A*, 986, 67–72 (2003).

60. Skoog, D. A., *Principles of Instrumental Analysis*, Sanders College Publishing, Philadelphia, PA, 1985.
61. Miyazaki, T., Yomota, C., and Okada, S., Ultrasonic depolymerization of hyaluronic acid, *Polymer Degradation Stabil.*, 74, 77–85 (2001).
62. Terbojevich, M., Cosani, A., and Palumbo, M., Molecular weight distribution of hyaluronic acid by high-performance gel-permeation chromatography, *Carbohydr. Res.*, 157, 269–272 (1986).
63. Lee, H. G. and Cowman, M. K., An agarose gel electrophoretic method for analysis of hyaluronan molecular weight distribution, *Anal. Biochem.*, 219, 278–287 (1994).
64. Nakano, T., Nakano, K., and Sim, J. S., A simple rapid method to estimate hyaluronic acid concentrations in rooster comb and wattle using cellulose acetate electrophoresis, *J. Agric. Food Chem.*, 42, 2766–2768 (1994).
65. Platzer, M., Ozegowski, J. H., and Neubert, R. H. H., Quantification of hyaluronan in pharmaceutical formulations using high performance capillary electrophoresis and the modified uronic acid carbazole reaction, *J. Pharm. Biomed. Anal.*, 21, 491–496 (1999).
66. Grimshaw, J., Kane, A., Douglas, A., Chakravarthy, U., and Archer, D., Quantitative analysis of hyaluronan in vitreous humor using capillary electrophoresis, *Electrophoresis*, 15, 936–940 (1994).
67. Bothner, H., Waaler, T., and Wik, O., Limiting viscosity number and weight average molecular weight of hyaluronate samples produced by heat degradation, *Int. J. Biol. Macromol.*, 10, 287–291 (1988).
68. Kachmar, J. F., Proteins and amino acids, in *Fundamentals of Clinical Chemistry*, Tietz, N. W., Ed., W. B. Saunders, Philadelphia, PA, 1987, pp. 177–262.
69. Lowry, O. H., Rosebrough, N. J., Farr, A. L., and Randall, R. J., Protein measurement with the folin phenol reagent, *J. Biol. Chem.*, 193, 265–275 (1951).
70. Uden, P. C. and Lavoie, L. M., Laboratory evaluation of commercial hyaluronate sodium products, *J. Equine Vet. Sci.*, 17, 123–125 (1997).
71. Swann, D. A., Studies on hyaluronic acid. II. The protein component(s) of rooster comb hyaluronic acid, *Biochim. Biophys. Acta*, 160, 96–105 (1968).
72. Murphy, D., Pennock, C. A., and London, K. J., Gas-liquid chromatographic measurement of glucosamine and galactosamine content of urinary glycosaminoglycans, *Clin. Chim. Acta*, 53, 145–152 (1974).
73. Balazs, E. A. and Sweeney, D. B., in *Replacement of the Vitreous Body of Monkeys with Reconstituted Vitreous and Hyaluronic Acid. Modern Problems in Ophthalmology (Surgery of Retinal Vascular Diseases and Prophylactic Treatment of Retinal Detachment*, Streiff, E. B., Ed., Karger, Basel, 1966, pp. 230–232.
74. Swann, D. A. and Constable, I. J., Vitreous structure. II. Role of hyaluronate, *Invest. Ophthalmol.*, 11, 164–168 (1972).
75. Constable, I. J. and Swann, D. A., Biological vitreous substitutes, *Arch. Ophthalmol.*, 88, 544–548 (1972).
76. Delinger, J. L. and Balazs, E. A., Replacement of the liquid vitreous with NaHA in monkeys. I. Short term evaluation, *Exp. Eye Res.*, 31, 81–99 (1980).
77. Delinger, J. L., El-Mofty, A. A. M., and Balazs, E. A., Replacement of the liquid vitreous with NaHA in monkeys. II. Long term evaluation, *Exp. Eye Res.*, 30, 101–117 (1980).
78. Balazs, E. A., Ultrapure hyaluronic acid and the use thereof, U.S. Patent 4,141,973, (1979).

79. McKnight, S. J., Giangiacomo, J., and Adelstein, E., Inflammatory response to viscoelastic materials, *Ophthalmic Surg.,* 18, 804–806 (1987).
80. Theodossiadis, G., Vitreous inflammatory reaction. Vitreous inflammation in rabbits, induced by saline, air, or sodium salt of hyaluronic acid, *Mod. Probl. Ophthalmol.,* 18, 265–270 (1977).
81. Cavaliere, S., Spampinato, D., Caruso, P., Dal Bello, A., Reibaldi, A., and Drago, F., Effect of hyaluronic acid fractions in the rabbit eye, *Ann. Ophthalmol.,* 22, 429–431 (1990).
82. Peiffer, R. L., The rabbit as an alternate model for the intraocular testing of viscoelastic substances, *Vet. Comp. Ophthalmol.,* 1, 83–86 (1991).
83. Shen, W.-Y., Constable, I. J., Celva, E., and Rakoczy, P. E., Inhibition of diclofenac formulated in hyaluronan on angiogenesis *in vitro* and its intraocular tolerance in the rabbit eye, *Graefes Arch. Clin. Exp. Ophthalmol.,* 238, 273–282 (2000).
84. Skelnik, D. L. and Lindstrom, R. L., Non-primate vitreal replacement model, U.S. Patent 5,102,653 (1992).
85. Floren, I., Viscoelastic purity, *J. Cataract Refract. Surg.,* 24, 145–146 (1998).
86. Wicken, A. J. and Knox, K. W., Lipoteichoic acids: a new class of bacterial antigen, *Science,* 187, 1161–1167 (1975).
87. Hogg, S. D. and Manning, J. E., Determination of *Viridans streptococci* surface lipoteichoic acid by enzyme linked immune sorbent assay, *FEMS Microbiol. Lett.,* 58, 239–242 (1989).
88. Tadler, M. B., Tice, G., and Wos, S. M., Sandwich immunoassay for the detection of lipoteichoic acid, *J. Clin. Lab. Anal.,* 3, 21–25 (1989).
89. Laurent, T. C., Dahl, I. M., Dahl, L. B., Engstrom-Laurent, A., Eriksson, S., Fraser, J. R., Granath, K. A., Laurent, C., Laurent, U. B., Lilja, K., The catabolic fate of hyaluronic acid, *Connect. Tissue Res.,* 15, 33–41 (1986).
90. Allen, S., Fraser, J. R., Laurent, U. B., Reed, R. K., and Laurent, T. C., Turnover of hyaluronan in the rabbit pleural space, *J. Appl. Physiol.,* 73, 1457–1460 (1992).
91. Fraser, J. R., Laurent, T. C., Engstrom-Laurent, A., and Laurent, U. G., Elimination of hyaluronic acid from the blood stream in the human, *Clin. Exp. Pharmacol. Physiol.,* 11, 17–25 (1984).
92. Smedsrod, B., Cellular events in the uptake and degradation of hyaluronan, *Adv. Drug Delivery Rev.,* 7, 265–278 (1991).
93. Fraser, J. R., Kimpton, W. G., Laurent, T. C., Cahill, R. N., and Vakakis, N., Uptake and degradation of hyaluronan in lymphatic tissue, *Biochem. J.,* 256, 153–158 (1988).
94. Reed, R. K. and Laurent, U. B., Turnover of hyaluronan in the microcirculation, *Am. Rev. Respir. Dis.,* 146, S37–S39 (1992).
95. Fraser, J. R., Laurent, T. C., Pertoft, H., and Baxter, E., Plasma clearance, tissue distribution, and metabolism of hyaluronic acid injected intravenously in the rabbit, *Biochem. J.,* 200, 415–424 (1981).
96. Baxter, E., Fraser, J. R., and Harris, G. S., Fractionation and recovery of secretions of synovial cells synthesized in culture with radioactive precursors, *Ann. Rheum. Dis.,* 32, 35–40 (1973).
97. Underhill, C. B. and Toole, B. P., Binding of hyaluronate to the surface of cultured cells, *J. Cell Biol.,* 82, 475–484 (1979).
98. Wilzbach, K. E., Tritium-labeling by exposure of organic compounds to tritium gas, *J. Am. Chem. Soc.,* 79, 1013 (1957).

99. Highsmith, S. and Chipman, D. M., Preparation of tritium-labeled hyaluronic acid oligomers and their use in enzyme studies, *Anal. Biochem.*, 61, 557–566 (1974).

100. Hook, M., Riesenfeld, J., and Lindahl, U. N-[3H]Acetyl-labeling, a convenient method for radiolabeling of glycosaminoglycans, *Anal. Biochem.*, 119, 236–245 (1982).

101. Swann, D. A., Silver, F. H., Sotman, S. L., and Hermann, H., Measurements of reducing end groups on bovine vitreous-humour hyaluronic acid by reaction with ^{14}C cyanide, *Biochem. J.*, 207, 409–414 (1982).

102. Orlando, P., Binaglia, L., De Feo, A., Orlando, M., Trenta, R., and Trevisi, R., An improved method for hyaluronic acid radioiodination, *J. Labelled Compounds Radiopharmaceut.*, 36, 855–859 (1995).

103. Gustafson, S., Bjorkman, T., and Westlin, J. E., Labelling of high molecular weight hyaluronan with 125I-tyrosine: studies *in vitro* and *in vivo* in the rat, *Glycoconj J.*, 11, 608–613 (1994).

104. Westerberg, G., Bergstrom, M., Gustafson, S., Lindqvist, U., Sundin, A., and Langstrom, B., Labelling of polysaccharides using [11C]cyanogen bromide. *In vivo* and *in vitro* evaluation of 11C-hyaluronan uptake kinetics, *Nucl. Med. Biol.*, 22, 251–256 (1995).

105. Orlando, M., De Sanctis, G., Giodano, C., Valle, G., La Bua, R., and Ghidoni, R., Tritium labeled hyaluronic acid derivative, *J. Labelled Compounds Radiopharmaceut.*, 22, 961–969 (1985).

106. Lindqvist, U., Westerberg, G., Bergstrom, M., Torsteindottir, I., Gustafson, S., Sundin, A., Loof, L., and Langstrom, B., [11C]Hyaluronan uptake with positron emission tomography in liver disease, *Eur. J. Clin. Invest.*, 30, 600–607 (2000).

107 Beaty, N. B., Tew, W. P., and Mello, R. J., Relative molecular weight and concentration determination of sodium hyaluronate solutions by gel-exclusion high-performance liquid chromatography. *Anal. Biochem.*, 147, 387–395 (1985).

5

Safety of Hyaluronan-Based Medical Products in Their Indicated Applications

Since the discovery of hyaluronic acid (HA), the understanding of its underlying scientific principles has continuously advanced. A great deal of information concerning the chemistry, physics, physiology, and so on is being generated through ever-increasing research activities in the U.S. and abroad. The convergence of the interest in basic research and the effort in product development led to series of HA products for a variety of medical applications.

The early 1980s ushered in the first U.S. Food and Drug Administration (FDA)-approved products containing HA as viscoelastic surgical adjunct in ophthalmic surgery (Healon[1] and Amvisc). Other medical applications of HA ensued in those areas of postsurgical adhesion prevention (Seprafilm[2], Intergel[3]) and the treatment of osteoarthritis of the knee through intraarticular injections (Synvisc,[4] Hyalgan,[5] Supartz[6,7]). More recently, HA products for wrinkle treatment (Restylane, Hylaform) also went through the FDA review and approval process. A list of the major HA-based products approved in the U.S. for medical use is provided in Table 1.1.

Chapter 5 focuses on the safety-related information and issues regarding HA products, starting from the safety related to HA products in general, followed by discussions of the safety profile of individual products as demonstrated in their clinical applications. Because the safety (and clinical performance) of a product should be viewed within the context of its actual indication and is better appreciated when the readers have the product description in mind (not just product names), Chapter 5 will provide more details of the specifications and indications of various HA products and will also organize the information in table form when appropriate (Tables 5.1–5.5).

5.1 General Safety Considerations

The compatibility of exogenously administered HA with the human body is an intrinsic property of HA. HA is a natural component of most human and animal tissues. The primary chemical structure of HA is identical, regardless of its source or molecular weight. The catabolism of exogenous HA is well known and remarkably rapid. These factors, in conjunction with the improvement in purification technology and analytical methods (chemical and biological), contribute to the safety profile of HA-based medical products that passed the scrutiny of the regulatory agencies in the U.S. and abroad.

5.1.1   HA — Natural Component of Human Tissues

The interest in the basic science of hyaluronan and the need for the development of HA-based products led to the intensive effort in the study of the catabolic fate of exogenous HA applied to human and animal bodies. HA is found extracellularly in most human tissues. The skin, lungs, and intestines contain more than 50% of the HA in the human body.[8,9] HA is also a rich component of synovial fluid, the umbilical cord, and the vitreous humor — the watery substance contained within the interstices of the stroma in the vitreous body of the eye.

The approximate weight of HA in 1 g of the above-referenced wet tissues are as follows: 0.5 mg in skin, 15–150 µg in lungs, 10–80 µg in intestine,[8] 0.2 mg in the vitreous humor of the eye, 1.4–3.6 mg in synovial fluid, and 3 mg in human umbilical cord.[10] This information indicates that the no effect level for exogenous HA may be greater than 3.6 g/kg of wet tissue.

5.1.2   Catabolic Pathways of HA

As mentioned in Chapter 1, the catabolic pathway of HA through lymphatic system, blood stream, and liver endothelial cells was established through the studies by Laurent et al.,[11] who injected radioactive HA into the knee joints of rabbits and traced the movement of its radioactivity. The study concluded that the HA released from tissue is mostly taken up and degraded by the lymphatic system, and once in the circulating system, it is bound, internalized, and catabolized by liver endothelial cells. Internalization of HA takes less than a minute.[11-13]

In the skin, where the concentration of HA is about 1 mg/ml,[8,14] HA is synthesized by both epidermal cells (keratinocytes) and the cells in dermis (principally fibroblasts).[15] The two compartments of the skins are separated by a basement membrane. Therefore, the clearance of HA in skin tissues is more complicated, and an effort has been made to investigate the specific contribution of each component to the overall catabolism.[15,16]

The turnover of HA in epidermis is also rapid, with its half-life being 1 day, during which the molecular weight of a newly synthesized HA was reduced from several million to half a million.[17] It is difficult for HA in the epidermis to cross the barrier of basement membrane and escape into the underlying connective tissue. There was evidence that the breakdown occurs locally in the epidermal tissue.[15] HA synthesized on epidermal keratinocytes is partially endocytosed and probably depolymerized by hyaluronidase and later completely degraded in lysosomes by exoglycosidases.[18]

Most of the HA in skins is sequestered within the dermal layer.[9,19] The removal rate of [³H] radiolabeled HA injected into the subcutaneous space was estimated to be 50–75% within 24 hours.[20] Dermal HA has access to the lymphatic and cardiovascular systems, where most of its degradation occurs.[19,21] There have been studies since the 1970s with regard to the possible existence of a local mode of HA turnover in dermis, in addition to the transport of HA via the lymphatic system mentioned earlier.[22,23] The efforts to identify the hyaluronidase activity associated with human dermal fibro-blasts were not successful until the studies by Stair-Nawy et al. in the late 1990s.[19] Note, however, that an active form of hyaluronidase *in vivo* within the extracellular matrix (ECM) is unlikely — otherwise, the very existence of a functioning ECM would be in question.

5.1.3 Physiological Turnover and Dose Tolerance

The rate of turnover of the exogenous HA and the residence time of the material *in vivo* are the "pharmacokinetic" aspects of the product safety. Related to that is the dose-dependent tolerance of HA-based products, which was investigated by administering the doses at and above the normal level of indicated use.

Only limited data are available from studies that apply radioactive HA to humans. Fraser et al. reported the study of the clearance of HA from the blood stream of four normal human subjects, who received intravenous injection of HA labeled with tritium in the acetyl position. The physiological turnover of HA in human bodies is remarkably rapid. The daily turnover of HA in the circulation system is in the range of 150–700 mg. The half-life of the injected HA in plasma ranged between 2.5 and 5.5 minutes. Rapid degradation was evidenced by the identification of ³H₂O in the urine of the subjects.[24]

Most of the information of clearance of HA has been from animal studies. In a rabbit study using IA injection of radiolabeled HA, according to the PMA (premarket approval) summary of Synvisc (*Summary of Safety and Effectiveness Data of Hylan G-F 20 P940015*, August 8, 1997), the half-life of one of its components, hylan A, was ~1.5 days, and for another component, hylan B gel, was ~8.8 days. As a comparison, the half-life of high–molecular weight native HA (molecular weight 1.7–2.6 million dalton) was 11 hours. The intravenous injection of Synvisc hylan G-F 20 was also attempted. The hylan A component was directly applied intravenously (half-life in the blood is 16–17 minutes), whereas hylan B was first reduced to a smaller polysaccharide because the insolubility and high viscoelasticity may cause the clogging of the blood vessel. The acid hydrolysis product of hylan B was considered to be "similar to the solubilized form of hylan B which exists in the joint and which enters the blood stream after hylan B is injected into the joint and degraded" (P940015 pp. 7–8). The half-life of hylan B in the blood by this method was 22 minutes. Other preclinical studies of Synvisc also concluded that the rate of clearance of Synvisc from blood of rat and rabbit was 15.9–77.4 mg/kg per day.

Synvisc was tested in owl monkey joints for its potential to induce local and systemic effects following weekly bilateral knee injections for 31 weeks. The animals received an amount of HA that represented "a dose greater than 20 times the total human clinical regimen when adjusted for body mass" (P940015, p. 6). During the observation period, local or systemic adverse effects attributable to Synvisc were not detected.

The absorption, distribution, and excretion of Intergel, ferric hyaluronate, were investigated in studies with rats, dogs, and monkeys. The serum levels of HA and iron were determined following intraperitoneal administration of Intergel. Slight increases in serum iron were observed within the first 48 hours following dose administration, but serum HA concentrations returned to near predose HA levels within 7 days. The studies concluded that "the elevated serum HA levels are transient and do not suggest accumulation at the clinical intended dose (5 ml/kg), or at doses up to 30 ml/kg" (P990015, p. 14). The half-life of Intergel in humans was estimated to be approximately 51 hours. With regard to dose tolerance, multiple-dose subchronic toxicity testing of Intergel resulted in no mortality for beagle dogs given up to 15 ml/kg per dose (7.5 mg/kg per dose) every third day for 28 days (*Summary of Safety and Effectiveness of Intergel, PMA P990015*, January 12, 2000).

The half-life of Hyalgan in synovial fluid was found to be approximately 17 hours according to a dog study. With regard to the dose tolerance of Hyalgan, an animal study was conducted in which Sprague-Dawley rats were administered 5, 10, or 20 mg/kg per day of Hyalectin (the equivalent of Hyalgan in Europe) by subcutaneous injection for 30 days. These amounted to a total dose of 150, 300, or 600 mg/kg body weight: "The cumulative exposure tested in this study is more than 100–400 times the anticipated clinical exposure dose of 1.43 mg/kg for one cycle of treatment" (P950027, p. 13). No adverse reactions were observed (*Summary of Safety and Effectiveness of Hyalgan P950027*, May 28, 1997).

More information with regard to Synvisc, Hyalgan, and Intergel can be found in Tables 5.2 and 5.3. It is noted that the characteristics of the three HA products mentioned above are, to a certain degree, representative of the HA-based medical products. Hyalgan is composed of native, unmodified HA. Synvisc is covalently crosslinked HA, and Intergel is an ionically crosslinked HA. Within the range of structural variations of these products, their turnover rates remain within an acceptable range for safety. Animal studies of Synvisc, Hyalgan, and Intergel have indicated that HA in the amount currently used in medical devices for therapeutic purposes does not accumulate significantly in the circulation, despite exposures hundreds of times greater than the recommended therapeutic use levels.

5.1.4   Consensus on Certain Specifications

To ensure that HA products are appropriate for their intended uses, various authoritative entities have established standards and guidance pertinent to the specifications for HA, as discussed in Chapter 4. Examples include those established by the European Pharmacopoeia for sodium hyaluronate powder and by the ISO for ophthalmic viscosurgical devices (OVDs).

The information concerning the exact specifications of individual medical products containing HA is generally regarded as proprietary and, therefore, not readily available to the public. However, most manufacturers, being attentive to the global harmonization trend and the requirement of local authorities, often follow those international standards conscientiously. Examples of requirements specified in these standards and guidance include those pertaining to pH, protein impurity, endotoxin level, and so on. All such standardized specifications, if followed, serve to ensure that any HA component of a device will not function to cause untoward effects in its indicated use, provided such use does not exceed that use level demonstrated to be safe.

5.1.5 Essential Safety Requirement in Preclinical Studies

Most HA-containing products are considered as implants and classified as medical devices. The *International Standard for Biocompatibility of Medical Device* ISO 10993 provides a universal guideline for the safety profile of such products and is widely applied in many preclinical studies. Over the last 20 years, the Center for Devices and Radiological Health (CDRH) of the FDA has reviewed and approved more than a dozen products containing HA. A list of those products and their indicated use and characteristics are summarized in Tables 5.1–5.5, as well as Table 1.1.

In most cases, preclinical study information was reviewed and found by CDRH to be sufficient to demonstrate that HA-based products were safe when used as directed. Such preclinical information indicated, among other things, that a purified HA is not mutagenic, pyrogenic, toxigenic, hemolytic, or immunogenic. It is not a dermal sensitizer, nor does it cause inflammatory response or damage tissue. This is consistent with the requirements of ISO 10993.

5.1.6 Effect of HA Chemical Modification on Its Safety

The structural change of HA through its chemical modification has been described in Chapter 2. It is well understood that the purpose of chemical modification of HA is to improve certain physical or biological properties of HA. What is yet to be fully appreciated by the general public is that a chemically modified HA, by definition and in fact, is chemically different from and less natural than the native HA. When dealing with chemically modified HA, the biocompatibility based on native HA cannot be taken for granted.

Depending on the degree and type of chemical modifications (including crosslinking) made, the physical properties of HA derivatives, such as solubility, viscosity, and elasticity, can be vastly different. The biological behavior of HA derivatives may turn out to be less compatible to human tissues than the native HA as well. The question is how much such change is acceptable in a specific medical application, weighed against the benefit the modification may bring.

5.2   Safety Information from Specific Human Use

In addition to the nature of the HA molecule and its product characteristics, the safety of HA-based products needs to be evaluated in the actual conditions of specific human use as well. The quality control measures, the *in vitro* and *in vivo* toxicity studies, and the other animal studies belong to the early-development stages of HA-based products, which would require the ultimate validation — human clinical experience.

The bulk of safety information of HA-based products in human use presented below is available through an FDA database and peer-reviewed articles and reports in medical journals. The FDA database is a rich source of information of product safety. In the U.S., HA-based medical products for human use are mostly classified as class III medical devices. One exception is ETX-100 of Exhale Therapeutics, Inc. (EXT-100 is an HA solution for nebulization to treat emphysema caused by cigarette smoking or alpha 1 antitrypsin deficiency. The compound was well tolerated in phase I trials in normal volunteers and in patients with antitrypsin deficiency. An orphan product designation has been obtained [see http://www.cotherix.com]). HA products for ophthalmic surgeries, postsurgical adhesion prevention, knee osteoarthritis, and cosmetic tissue augmentation have gone through a stringent application/approval process (PMA), and HA products used as nasal dressing or otologic pack have been approved through a simpler regulatory process — 510(k) — which only needs to demonstrate equivalency with a predicate product legally marketed before May 28, 1976. In the case of PMAs, detailed, useful information with regard to product safety, as demonstrated in human clinical trials, has been made available by the FDA through the publication of certain regulatory documents, and in particular the *Summary of the Safety and Effectiveness* of the approved products. Once the products enter the market, the MAUDE (Manufacturer and User Device Experience) database of FDA/CDRH begins to provide further information regarding the safety of the products in the form of voluntary reports of adverse events filed by health care professionals, distributors, or the manufacturer as they are generated. The information is basically a narrative of the events by the reporters, however, and could be anecdotal.

5.2.1   Ophthalmic Surgical Aid

The major HA-based products approved by the FDA as medical devices for ophthalmic surgeries are listed in Table 5.1. The information regarding product features, indications, and manufacturers is in accordance with the regulatory approval documents unless otherwise specified.

The safety of a specific product should be viewed in the actual conditions of its applications, and evaluated based on the relative benefit and risk factors (*Determination of Safety and Effectiveness* [21 CFR § 860.7(d)(1)]: "There is reasonable assurance that a device is safe when it can be determined, based upon

TABLE 5.1

HA Products Approved for Marketing in the U.S. for Ophthalmic, Surgical Use

Product and PMA Number	Product Description	Indications

HEALON
 	
P810031
	HA source: rooster combs
	Media: PBS
	MW Healon: 2–4 million
	        Healon GV: 5 million
	        Healon5: 4 million
	[HA] Healon: 10 mg/ml
	        Healon GV: 14 mg/ml
	        Healon5: 23 mg/ml
		Healon as surgical aid in cataract extraction, IOL implantation, corneal transplant, glaucoma filtration, and retinal attachment Healon GV and Healon5 in particular create and maintain a deep anterior chamber, and facilitate manipulation inside the eye with reduced trauma to the ocular tissues.

AMVISC
 	
P810025
	HA source: rooster combs
	Media: physiological saline
	MW: Amvisc ~ 2 million
	      Amvisc Plus: 1.5 million
	[HA] Amvisc: 12 mg/ml
	      Amvisc Plus: 16 mg/ml
		Surgical aid for cataract extraction, IOL implantation, corneal transplant, glaucoma filtration, and retinal attachment

VISCOAT P840064	A mixture of HA and CS CS source: shark Media: PBS MW: HA > 500,000 CS: ~ 55,000 [HA] ~ 30 mg/ml [CS] ~ 40 mg/ml	Surgical aid for anterior segment procedures including cataract extraction and IOL implantation.
PROVISC P890047	HA source: fermentation Media: PBS MW: not specified [HA] 10 mg/ml	Surgical aid for anterior segment procedures including cataract extraction and IOL implantation
VITRAX P880031	HA source: biological Media: PBS MW: not specified [HA] 30 mg/ml	Surgical aid for anterior segment procedures including cataract extraction and IOL implantation, corneal transplant and glaucoma surgery
BIOLON P960011	HA source: fermentation Media: PBS MW: 3 million [HA] 10 mg/ml	Surgical aid for cataract extraction and IOL implantation; maintains a deep anterior chamber.

CS = chondroitin sulfate, [CS] = CS concentration, [HA] = HA concentration, IOL = intraocular lens, MW = molecular weight, PBS = phosphate buffered solution.

valid scientific evidence, that the probable benefits to health from use of the device for its intended uses and conditions of use, when accompanied by adequate directions and warnings against unsafe use, outweigh any probable risks"). Therefore, we first need to refer to the intended use and actions of those products in ophthalmic surgeries. In cataract extraction surgery, HA is used to maintain a deep anterior chamber. In IOL implantation, HA protects the corneal endothelium and other ocular structures. In corneal transplant, HA aids in the placement, suturing, and maintenance of the shape of the donor graft. In glaucoma filtration surgery, HA reduces the incidence and severity of postoperative anterior shallowing. In retinal attachment surgery, HA can serve as a vitreous placement and hold the retina against the sclera for reattachment, at least temporarily (PMA: P810031, P810025, P840064).

The safety of HA products is often determined, at least in the premarketing stage, by comparing it with alternatives that were already on the market and deemed safe. Air or other gases and isotonic solutions such as Ringer's lactate solutions or balanced salt solution were considered to be such alternatives. They constitute the "commonly used articles for restoration of the volume of the eye" (P810031).

Healon was the first HA-based viscoelastic product for ophthalmic use to receive approval from the FDA (the approval dates of HA products are listed in Table 1.1). Certain safety parameters of Healon were examined and later applied to other HA products with similar indications. For instance, intraocular pressure (IOP) was monitored because of the concern that increased IOP might be associated with Healon and that the increases in IOP could be sight threatening. It was thought that Healon, as a high–molecular weight HA formulation, may be slow in exiting the eye after surgery, and even blocking the trabecular meshwork. The preapproval clinical studies demonstrated that such postoperative elevations in IOP were "generally mild, occurred in the early postoperative period, and resolved spontaneously or responded to conventional therapy" (P810031).

Another safety indicator was the endothelial cell count. Healon appeared to show a trend of less endothelial cell loss than the BSS and Ringer's lactate solutions that the investigators used in the studies as controls.

In summary, the FDA concluded that the adverse reactions encountered during the surgery were not attributed to Healon, and the collective data from clinical studies constituted valid scientific evidence of the safety of Healon for the proposed indication. Amvisc showed a similar safety profile as described in its PMA summary (P810025): "There were no reports of intraocular infection, severe inflammatory response, or any adverse reactions in any patient receiving Amvisc. No problems other than the reported mild, transient inflammatory response were encountered during surgery that could be attributed to Amvisc." Viscoat is a combination product composed of HA and another glycosaminoglycans; namely, chondroitin sulfate. Safety data were collected during two postoperative visits, during 1–3 days for the first visit and 4–15 days for the second visit. There was no statistically significant difference in the rate of clear cornea postoperatively between Viscoat and the controls (historical data), although the result appeared to be in favor of Viscoat numerically in both visits. The FDA allowed the use of data from another prospective clinical study as control, because the "parameters measured were identical to those evaluated in the Viscoat study." The control patients in other studies received identical types of anterior segment surgery except, rather than receive Viscoat, they received balanced salt solution or air or no injection of any substance into the anterior chamber (P840064). In addition, the rate of cornea edema for Viscoat was significantly lower than control during the second visit. Endothelial cell loss for Viscoat group was 8.04% — lower than the results in the medical literature for conventional surgery.

The rate of postoperative IOP equal to or greater than 20 mmHg in the Viscoat group was higher (23.7%) than the control (14.2%) during the first visit, with statistical significance. During the second visit, the rate in the Viscoat group was also higher (12.1%) than the control (8.7%) but was not statistically significant. The increased IOP in both cases, however, was transient and controllable.

Viscoat in its earlier formulation was found to form a white precipitate in the cornea stroma postoperatively at an incidence of less than 1%. The precipitate formations did not adversely affect visual acuity. Nevertheless, the manufacturer identified the cause of the precipitate as calcium phosphate and later reduced the concentration of phosphate. The follow-up clinical study using the new formula found no evidence of precipitate formation.

The safety profiles of Vitrax, Provisc, and Biolon demonstrated similar safety patterns in terms of endothelial cell loss, cornea edema, and increase of IOP. They were compared to those HA products previously approved for ophthalmic surgery as well as salines. For instance, during the PMA review of Provisc, the other HA-based OVDs such as Healon, Amvisc, and Viscoat were listed by the FDA as alternative practices, along with balanced salt solution and Ringer's lactate solutions.

It is noted that sometimes the distinction between safety and effectiveness may not be absolutely clear. For example, in the *Summary of Safety and Effectiveness Data of Provisc*, endothelial cell density, corneal thickness, and corneal edema were grouped into efficacy data, instead of being under safety (P890047).

The deficiency or anomalies in manufacturing and quality control affecting the purity of the products are always a safety concern and need to be monitored closely. According to Floren, several cases of endophthalmitis-like reactions occurred after using Microvisc in cataract surgeries in 1997. The Swedish regulatory authorities tested Microvisc and found endotoxin content as high as 100 EU/ml in multiple samples, and they withdrew Microvisc from the market immediately.[25] The problem of high endotoxin counts with Microvisc later appeared to be resolved, according to a report by Dick et al. in 2003.[26] The study measured the endotoxin level of 25 commercially available OVDs, and the endotoxin content of Microvisc was found to be "under 1.2 endotoxin unit/ml."[26]

Close attention should also be paid to the indication and use of the product as directed by the package insert. For example, during vitrectomy in aphakic eyes, "Healon diffused rapidly into the anterior chamber, . . . causing marked elevations of intraocular pressure."[27] In another case, the intravitreal-injected NaHA caused persistent intraocular pressure rise to 60–70 mm Hg in three aphakic eyes, which "necessitated removal of Healon to normalize the intraocular pressure."[28] Accordingly, the package insert of Healon included the following precaution: "in posterior segment procedures in aphakic diabetic patients special care should be exercised to avoid using large amounts of Healon" (Healon Package insert, http://www.healon.com).

In short, the safety data of the HA-containing products such as OVDs indicate that HA products, formulated within the defined molecular weight, concentration, osmotic pressure, and pH, are safe, provided they are thoroughly purified and used as directed.

5.2.2   Intraarticular Injection in the Knee

The features and indications of four of the HA-based products approved by the FDA for intraarticular injection to treat the pain of osteoarthritic knees are listed in Table 5.2. They are Hyalgan, Synvisc, Supartz, and Orthovisc. Again, keep in mind that the safety of a product should always be examined within the context of its intended use, and that a product is by and large defined by its composition and formulation characteristics.

TABLE 5.2

HA Products Approved for Marketing in the U.S. for Intraarticular Injection of the Knee

Product and PMA Number	Product Description	Indication and Doses
SYNVISC P940015	Elastoviscous liquid of hylan (crosslinked HA by divinylsulfone and formaldehyde) HA source: rooster combs Media: PBS (pH 7.2) MW ~ 6 million for hylan A [HA] 16 mg/2ml	Treatment of pain of OA knee in patients who failed to respond adequately to conservative therapies including acetaminophen. 3 weekly IA injections
HYALGAN P950027	Viscous solution of purified native HA HA source: rooster combs Media: PBS (pH 6.8–7.5) MW: 500,000–730,000 [HA] 20 mg/2ml	Same indication as Synvisc 5 weekly IA injections
SUPARTZ P980044	Viscous solution of purified native HA HA source: chicken combs Media: PBS (pH 6.8–7.8) MW: 620,000–1,170,000 [HA] 25 mg/2.5ml	Same indication as Synvisc 5 weekly IA injections
ORTHOVISC P030019	Viscous solution of purified native HA HA source: rooster combs Media: PBS (pH 6.8–7.5) MW: 1.0–2.9 million [HA] 30 mg/2ml	Same indication as Synvisc 3 or 4 weekly IA injections

[HA] = concentration of HA or its derivatives, IA = intraarticular, MW = molecular weight, OA= osteoarthritis, PBS = phosphate buffered solution.

5.2.2.1   Clinical Information in PMA Documents Published by the FDA

The clinical trials of Hyalgan, Synvisc, Supartz and so on generated much information with regard to the safety of those products. The information was summarized by the sponsors of those studies and submitted to the FDA for review before an approval for marketing could be granted. In addition to the data from the biocompatibility and other preclinical studies, adverse events in those human clinical trials were analyzed. In all those studies, adverse events were found to be mild, transient, and not product related, as compared to the controls — buffered saline. In some studies, control groups also included oral administration of drugs for OA, such as Naproxen, a nonsteroid antiinflammatory drug, as described in the following paragraphs.

5.2.2.1.1   Hyalgan

The pivotal study of Hyalgan submitted for FDA review enrolled a total of 495 patients, randomly assigned to three groups: HA injection/placebo pill, saline injection/placebo pill, and sham injection/Naproxen. In that study, the adverse events more frequently encountered (>10%) were reported as gastrointestinal complaints, pain at injection site, headache, and local joint pain and swelling (see Table 1 of *Summary of Safety and Effectiveness of PMA 950027*). Most of the differences between groups were not statistically significant, except for two instances. First, the Naproxen group had more gastrointestinal complaints (41%) than the Hyalgan group (29%). This was probably anticipated, as gastrointestinal toxicity was, after all, a major known side effect of nonsteroid antiinflammatory drugs and one of the reasons why HA injection products were developed as an alternative in the first place. Second, it was noted that Hyalgan group had higher incidence of pain at injection site (23%) than the saline injection group (13%). The reason for that difference was not exactly clear. Four cases of severe knee swelling and effusion and five cases of infections were considered as severe adverse events. They were low-incidence events (1–2%) and were evenly distributed among the study arms.

5.2.2.1.2   Synvisc

A series of Synvisc studies served as the basis for FDA approval, which included seven clinical trials in three countries: the U.S., Germany, and Canada (*Summary of Safety and Effectiveness Data of P940015*). The studies enrolled a total of 511 patients who received 1771 injections. Local adverse events such as knee pain or swelling occurred in 7% of patients (2.3% of injections). Systemic adverse events occurred in 10 patients (2%), with cases of rash and itching of the skin and so on. It was concluded, however, that "there were no statistically significant difference in the number or types of adverse events between the group of subjects that received Synvisc and the group that received buffered saline as control treatment."

5.2.2.1.3   Supartz

Supartz studies (*Summary of Safety and Effectiveness Data of P980044*) included five saline-controlled clinical trials, in which 619 patients received Supartz and 537 received control injections. Note the two most common adverse events in the studies, which occurred in greater than 10% patients, were arthralgia and arthropathy/arthrosis/arthritis. Arthralgia was defined here as joint pain with no evidence of inflammation (17.8% for Supartz and 17.7% for saline control), and arthropathy/arthrosis/arthritis was defined as joint pain with evidence of inflammation (11.0% for Supartz and 10.6% for control). The incidence rates between treatment groups were not statistically significant.

As a common phenomenon observed in the three studies, the transient, local pain and swelling caused by the injection procedure and the injected material were the major adverse effect of such therapy. This side effect should be taken into consideration when weighing the balance between the risk and benefit of such therapy. However, the right injection technique and the experience of clinicians can play a critical role in the success of such therapy. Jones et al. studied the accuracy of intraarticular injections by rheumatologists using radio contrast material and found the injected material outside the joint space of approximately 30% of the patients.[29] Given the high frequency of mis-placement of the injected material, additional measures of caution by clini-cians may be warranted, such as a rapid, on-site assay to detect the presence of synovial fluid in the joint before injection.[30]

Needless to say, all FDA-approved HA-based products were deemed safe at the time of approval, on the basis of the totality of evidence of safety as determined by the agency. However, product-related adverse events, whether pre– or post–marketing approval, are relevant safety concerns and should be addressed. It may be required that the original labeling of the approved product be revised to include the postmarketing clinical data or the new perspective of the old data in light of the new clinical experience. A case to the point is the revision of the package insert of Synvisc for "more fully describing the U.S. clinical trial, describing the signs of an allergic reaction," as indicated in a 2003 supplement to the PMA of Synvisc (P940015/S010, Synvisc® Hylan G-F 20, September 12, 2003).

5.2.2.2   Adverse Events Reported in Medical Journals

The human clinical experience of HA-based products for osteoarthritis is, of course, not restricted to those studies mentioned above. It has been over 7 years since the approval of Synvisc and Hyalgan by the FDA in 1997, and the marketing history of those products outside the U.S. can be traced back even earlier. Synvisc started marketing in Canada in 1992 and Hyalgan in Europe in 1987. Supartz was approved by the FDA in 2001. Before that, Supartz (Artz) already had a marketing history in 11 other countries.

As the market grows, new data with regard to the safety of those products continue to emerge in medical journals and the FDA/CDRH database, MAUDE. On the basis of such information, Hamburger et al. wrote a review of product-specific safety profiles of Hyalgan, Synvisc, and Supartz.[31] The paper concluded that overall, HA therapy is a safe treatment for OA knee pain. However, one HA product, "Synvisc, has been associated with a disproportionately high frequency of SAIRs" (severe acute inflammatory reactions).

Buchner et al. characterized SAIRs or pseudoseptic reactions as "severe pain, warmth, and swelling, associated with an effusion with cellular infiltrate (frequently monocytes and/or eosinophils) that may last up to 3 weeks without clinical intervention, . . . and by synovial fluid effusions that are negative on culture (ruling out septic arthritis) and devoid of calcium pyrophosphate crystals (ruling out pseudogout)." Prior injection is also generally required.[32] A distinction was thus made between SAIRs and other inflammatory reactions.

5.2.2.2.1 Pseudogout and Similar Cases

In 1999, Ali et al. reported a case of acute pseudogout of a patient following the third injection of Synvisc. Microscopic analysis revealed multiple intracellular rhomboid crystals typical of pseudogout.[33] Yacyshyn and Matteson reported a case of gout. The patient had received a first series of three injections of Synvisc 8 months earlier and experienced symptomatic improvement but developed severe pain and bilateral knee swelling 6 hours after the first injection of the second series. "Synovial fluid obtained at arthrocentesis revealed evidence of intraarticular uric acid crystals."[34] In 2000, Kroesen et al. reported a case of acute calcium pyrophosphate dihydrate arthritis. A patient developed pain and swelling of the knee 2 days after the second injection of Synvisc. There was also severe loss of physical function. Although calcium pyrophosphate dihydrate crystals were found in synovial fluid under microscope, bacterial contamination was not detected.[35]

The occurrence of pseudogout following intraarticular injection has not been restricted to Synvisc. In 1998, Luzar and Altawil reported a case of acute attack of pseudogout arthritis immediately following Hyalgan injection.[36] Maillefert et al. in France reported two similar cases a year before.[37]

5.2.2.2.2 SAIRs (Pseudoseptic)

Puttick et al. in Canada reported acute local reactions after intraarticular injection of Synvisc into the knee. Among 22 patients (88 injections to 28 knees), 6 patients had reactions within 24 hours of injection, characterized by pain, warmth, and swelling, and lasting up to 3 weeks. Synovial fluid cell counts were $5.0-75.0 \times 10^9$/L, often with a prominent mononuclear component. Crystal studies and cultures were negative. One patient had serum antibodies to chicken serum proteins.[38]

Allen and Krohn reported a single case of allergic reaction of Synvisc, which occurred after the third injection. The first two injections were well tolerated. However, within 2 hours of the third injection, the patient rapidly developed painful swelling of the knee, and they developed a very large effusion in less than 24 hours. "There were no crystals. Gram stain revealed leukocytes but no bacteria. Cultures of both synovial fluid and peripheral blood were negative at 72 hours."[39]

Martens reported a case of synovitis that occurred following injections of Synvisc. The patient reaction was characterized by swollen and painful knee, with a large volume of effusion that was grossly turbid.[40] Rees and Wojtulewski in the United Kingdom also reported a case of severe reaction to Synvisc that appeared to fit the same pattern. The first two injections had no complications — the severe reaction occurred 12 hours after the third injection.[41]

Leopold et al. investigated the frequency of painful acute local reactions following injections of Synvisc. They found that such reactions occurred significantly ($P = .029$) more often in patients who had received more than one course of treatment (21%) than in patients who had received only a single course of treatment (2%). All of the reactions were severe enough to cause the patient to seek unscheduled care but abated following corticosteroid injection.[42]

Pullman-Mooar et al. evaluated eight patients with painful local reactions after intraarticular injection of Synvisc and found that none of those flares followed the first injection, and only one seemed to be related to crystals. This indicates that sensitization to or accumulation of Synvisc or its breakdown products may play an etiologic role in these flares.[43] No detrimental lasting results were noted.

5.2.2.2.3 Foreign-Body Reaction

Zardawi and Chan in Australia reported a case of perisynovitis caused by the injection of Synvisc into the perisynovial adipose tissue. It was purportedly the "first histological account of tissue reaction to Synvisc." The report concluded, "injection of Synvisc into the synovial membrane or the paraarticular tissues can incite a granulomatous reaction, with subsequent scarring and limitation of movement."[44]

Although rare, there have been reports of cases of severe and long-lasting adverse effects that may require surgical intervention. Chen et al. observed chronically inflamed synovium with areas of histiocytic and foreign-body giant-cell reaction surrounding a cellular, amorphous material in patients treated with Synvisc. The five patients treated with Synvisc in the study actually underwent a surgical procedure because of persistent symptoms. The material stained with Alcian blue and disappeared after hyaluronidase digestion. This strongly indicates that the injected Synvisc may have been a pathological cause of recalcitrant symptoms after such injection, although it was not clear whether the responsible pathological agent was the HA derivative or a contaminant in the formulation.[45]

Other clinicians echoed the sentiment expressed by Chen et al. to raise clinical awareness of this potential complication. Mont and Etienne, in a letter commenting on the report by Chen et al., pointed out that they had not seen the type of inflammatory reaction to the Hyalgan — an "un-modified hyaluronate-derived preparation." They further hypothesized that the "chemical crosslinking of Synvisc may enhance the immunogenicity of components"[46] Morton and Shannon concurred and stated that they administered Hyalgan to patients having reactions to Synvisc, "with good clinical benefit and without further sequelae."[47]

5.2.2.2.4 Is Synvisc Nonimmunogenic?

The hypothesis that immunologic sensitization is involved in some of the adverse reactions of Synvisc prompted Bucher et al. to test directly the immunogenicity of HA preparations in animals. In an animal study, all four rabbits immunized with the positive control and three of the four rabbits immunized with Synvisc exhibited an anti–chicken protein response. None of the four rabbits immunized with Hyalgan had a detectable response to chicken protein.[32]

Sasaki et al. of Seikagaku Corporation, Japan, compared Supartz and Synvisc and found that Synvisc (hylan G-F 20) induces delayed foreign-body inflammation in guinea pigs and rabbits. When compared to saline and natural HA, Synvisc induced definitive macroscopic changes in guinea pigs by day 14 after intradermal injection, and in rabbits by 28 days postintramuscular injection. Furthermore, "specific antibodies against hylan were demonstrated in guinea pigs by passive cutaneous anaphylaxis, and substantial deposits of IgG on hylan were evident by immunohistochemistry."[48]

5.2.2.3 Postmarketing Safety Information from MAUDE

According to the FDA/CDRH MAUDE database, which contains data received through December 29, 2004, there had been 87 adverse events (AEs) reported for Hyalgan (since 1997.9), 271 AEs for Synvisc (since 1998.6), and 33 AEs for Supartz (since 2001.7). In the FDA/CDRH MAUDE database in September 2003, the numbers of corresponding AEs were 64, 142, and 24 respectively. These numbers are rather small considering that total injections in the knee of such products in the U.S. are estimated to be in the millions ("More than 102 million Supartz injections have been sold worldwide since its introduction in 1987"; see http://www.smith-nephew.com/investors/portfolio/ortho-Supartz.html. The total sale of Synvisc in 2003 exceeded 100 million U.S. dollars [Genzyme 2003 Annual Report]). This is, in general, consistent with the premarketing conclusion that all three products are basically safe.

However, reports from medical journals described above indicate that product-specific adverse events in the form of severe acute inflammatory reactions do exist. Hamburger analyzed the MAUDE data of Hyalgan and Synvisc from June 1998 to March 2002 and found 26 SAIRs or possible SAIRs (39% of total 67 adverse events) related to Synvisc, but only 6 SAIRs or possible SAIRs (19% of total 32 AE) related to Hyalgan.[31] This appears to be consistent with the information published in medical journals mentioned earlier.

5.2.3   Prevention of Postsurgical Adhesions

Two HA-based products, Seprafilm and Intergel, have been approved by the FDA so far for use in OB-GYN and abdominal surgeries to prevent or reduce the rate of postsurgical adhesions. Seprafilm is a bioabsorbable membrane (the name was changed from Seprafilm Bioresorbable membrane to Seprafilm Adhesion Barrier on December 20, 2000; P950034/S017), and Intergel is a liquid (Table 5.3). Both are chemically modified. Much information regarding their safety has been available through the publication of the regulatory review documents. However, the postmarketing information from articles in medical and other professional journals is limited as compared to that of the use of HA for knee injection mentioned earlier.

5.2.3.1   Safety of Seprafilm

Data from clinical studies of Seprafilm were submitted to CDRH of FDA for review (*Summary of Safety and Effectiveness Data of Seprafilm, P950034*). A total of 172 patients received Seprafilm, and the control groups received no treatment material. The two studies devoted to safety (HF91-1201 for abdominal surgeries and HF91-1202 for gynecological surgeries) enrolled only 15 and 17 patients, respectively. Although the numbers were quite small, they gave some preliminary information of the safety of the product in terms of adverse events such as fever and vital signs, as well as some baseline laboratory values.

A more definitive conclusion could be drawn from the other two studies with larger patient populations. HF92-0901 was a study of Seprafilm in abdominal surgery with 183 patients (91 Seprafilm, 92 no treatment), and

TABLE 5.3

HA Products Approved for Marketing in the U.S. for the Prevention of Postsurgical (Abdominal, OB-GYN) Adhesions

Product and PMA Number	Product Description	Indications
SEPRAFILM (Adhesion Barrier) P950034	Bioabsorbable membrane composed of CDI-modified HA and CMC HA source: fermentation Size: 3″ × 5″ and 5″ × 6″	For patients undergoing abdominal or pelvic laparotomy to reduce the incidence, extent and severity of PSA between abdominal wall and underlying viscera.
INTERGEL (Gynecare Intergel Adhesion Prevention Solution) P990015	A viscous, amber-colored solution composed of HA crosslinked with ferric ions. Media: PBS (pH 6.8–7.5) [HA] 5 mg/ml Unit volume: 300 ml	For patients undergoing open, conservative gynecologic surgery to reduce PSA and to reduce the likelihood of developing moderate or severe postoperative adnexal adhesions.

CDI = carbodiimide, CMC = carboxymethylcellulose, PSA = postsurgical adhesion, PBS = phosphate buffered solution, [HA] = concentration of HA or its derivatives

HF92-0902 was a gynecological study with 127 patients (59 Seprafilm, 68 no treatment). Over 90% of the patients of each group in both studies experienced one or more adverse events, mostly reported as nausea, abdominal pain, and fever, and the between-group difference was small. This indicates that the events were more surgery related than Seprafilm related.

In the abdominal surgery study (HF92-0901), 35% of the treated patients and 23% of the control patients experienced serious adverse events, which approached statistical significance ($P = .074$) between groups. It was noted that abscesses were four times more frequent in the Seprafilm group (8%) than in the control group (2%). Two cases of deep vein thrombosis and four cases of pulmonary embolus were reported for the treatment group, and only one thrombosis and no pulmonary emboli were reported for the control group. In the gynecological study (HF92-0902), only two patients (3.4%) of the Seprafilm group and three patients (4.4%) of the control group had serious adverse events (ileus and fever), with no incidence of abscess, pulmonary embolus, or thromboembolic events in that study.

The amount of material used for medical application is a factor affecting product safety. In the ophthalmic applications described earlier, HA (as a surgical aid) is meant to be removed after surgery. For intraarticular injection, patients receive exactly the same dosing, as the concentration and volume of HA and times of the HA injection are clearly defined. For the use of Seprafilm, each patient may receive a different amount of material, and therefore only average quantities can be determined. For example, for patients undergoing major abdominal surgery (HF-92-0901), the mean total quality of Seprafilm membrane (5" × 6") applied in the active group was 2.3 membranes per patient.

Beck et al. conducted a postmarket study of Seprafilm, which was a prospective, randomized, multicenter, multinational, single-blind, controlled study of 1791 patients undergoing abdominopelvic surgery.[49] Patients received an average of 4.4 pieces of Seprafilm. The barrier product was applied to the tissue surfaces that sustained direct surgical trauma and to suspected adhesiogenic sites as well. The number of pieces of Seprafilm applied was as high as 10. No foreign-body reaction was observed in the study. The difference for the incidence of abscess between the Seprafilm (4%) and control groups (3%) was not statistically significant ($P > .05$).

In Beck's study, however, fistula and peritonitis occurred more frequently in the Seprafilm group (~2%) than the control (<1%), with statistical significance ($P \le .05$). This may be attributed to a subpopulation of patients who underwent bowel anastomosis and had Seprafilm wrapped around the fresh anastomosis site. For those patients, leak-related events (including anastomotic leak, fistula, peritonitis, abscess, and sepsis) occurred more frequently ($P \le .05$). Therefore, the paper warned against "wrapping the suture or staple line of a fresh bowel anastomosis with Seprafilm." The above results from the Beck study underscore the importance of examining the potential effect of biomaterials on wound healing when a safety issue is being addressed.

In fact, FDA guidance for such biomaterials (*FDA Guidance for Resorbable Adhesion Barrier Devices for Use in Abdominal and/or Pelvic Surgery*; Guidance for Industry Document issued on June 18, 2002) has included "Testing for Delay of or Prevention of Healing."

5.2.3.2   Safety of Intergel

The product approval of Intergel in the U.S. went through a quite unusual process (*Summary of Safety and Effectiveness Data of PMA P990015 of Intergel*). In January 2000, the General and Plastic Surgery Devices Panel of the FDA reviewed the results from the clinical studies of Intergel used in laparotomy for gynecologic interventions. The panel made the initial recommendation that the PMA of Intergel was not approvable, mostly for safety concerns of the potential contribution of the device to the infection rate. The infection rate for Intergel was 3.9%, and the rate for control (lactate Ringer's solution) was 1%. The data represented clean and nonmalignant cases, whereas cases subject to contaminations, such as the opening of gastrointestinal or genito-urinary tract, or removal of an anatomic site, were excluded (Dr. Roxi Horbowyj, presenting the FDA clinical review on Intergel, FDA Medical Devices Advisory Committee, General and Plastic Surgery Devices Panel, January 12, 2000).

In June 2000, Lifecore amended the PMA of Intergel with modified indications for conservative gynecologic pelvic surgery — an obvious attempt to increase its chance of approval. The agency, however, came to the conclusion in November 2000 that the PMA did not provide reasonable assurance that the product was safe and effective, and therefore it was not approvable. In January 2001, Lifecore requested that the PMA be further reviewed by the Medical Devices Dispute Resolution Panel. In the subsequent review of the PMA in September 2001, the panel "had fewer questions about the safety of Intergel and discounted the importance of the trend that suggested a slightly higher infection risk for those receiving Intergel." The mitigated safety concern tipped the balance of opinion in favor of the benefit of the product, which was considered by some panel members as modest clinically, but better than no therapy. The panel recommended during the meeting that PMA of Intergel be approved without conditions. Lifecore gained formal approval from FDA to market Intergel in November 2001. Intergel was indicated for use in "open, conservative gynecological surgery as an adjunct to good surgical technique to reduce post-surgical adhesions."

The marketing of Intergel made a dramatic turn when its distributor, Gynecare Worldwide, a division of Ethicon, Inc., decided to withdraw the product from the global market voluntarily in April 2003. The action was taken in recognition of the postmarketing adverse events with the device believed to be associated with "off-label use in laparoscopy and non-conservative surgical procedures such as hysterectomy."

The announcement for withdrawal of Intergel was posted on Med-Watch, the site for the FDA Safety Information and Adverse Events Reporting Program.

The adverse events include late-onset postoperative pain that could be indicative of other serious complications of which physicians should be aware. Gynecare stated, "from the launch of this device in 1998 to February 2003, the overall complaint rate worldwide is low," and intended to conduct a "full and thorough assessment of technical issues, surgical techniques and circumstances associated with the post-market events" (http://www.fda.gov/medwatch/SAFETY/2003/Intergel.htm; "Urgent Global Market Withdrawal: Gynecare Intergel Adhesion Prevention Solution Voluntarily Withdrawn from the Market" by Gynecare Worldwide, April 16, 2003).

Postmarketing safety information from the MAUDE database appears to corroborate the above statement that many of the adverse events can probably be attributed to off-label use. From February 22, 2002, to December 20, 2004, a total of 187 cases of adverse events of using Intergel were reported in the MAUDE database. A search for published journal articles analyzing such data has yielded no results.

The clinical experience of Intergel is an example of the large effect that an off-label use may have on a medical product in a real world. The likelihood of a product to be used beyond the scope defined in the label tends to increase as the indication gets narrower. Therefore, the potential for off-label use and its consequences should be part of the product design considerations. In addition, the safety and effectiveness are often interrelated. The need to use a large volume of HA and to keep it in the abdominal and pelvic cavity postsurgery is a result of the marginal effectiveness of the product, which will be discussed in more detail in Chapter 6.

5.2.3.3 *Infectivity Issue*

The 2002 "Guidance for Resorbable Adhesion Barrier for Abdominal and Pelvic Surgery" of the FDA specially addressed the need to test the infectivity of material for such applications. The safety related to the adhesion prevention products is complicated by the fact that these products are often used in an environment in which surgical manipulation may bring accidental contamination, as in the case of bowel disruption. As noted in the Seprafilm package insert, "the safety and effectiveness of Seprafilm adhesion barrier has not been evaluated in clinical studies in the presence of frank infections in the abdominopelvic cavity" (Seprafilm Adhesion Barrier, Genzyme Biosurgery, 2004). In the case of Intergel, as mentioned earlier, the higher incidence of infection was one of the reasons for disapproval by the FDA in 2000. When the FDA panel reviewed the case for the second time in 2001, it discounted the importance of a slightly higher infection risk from Intergel. The issue of infectivity, however, did not go away, especially in light of the serious side effects allegedly associated with the off-label use.

Animal studies on the infectivity conducted after the approval of Seprafilm in the U.S. have been reported with some interesting results. Tzianabos et al.[50] conducted an intraabdominal infection study of four materials in an experimental rat model:

1. Seprafilm, (HA and CMC chemically modified by the carbodiimide EDC
2. a gel made of the same chemically modified material
3. an HA gel with ferric ions (mimicking the formulation of Intergel)
4. a saline control

Inoculum of bacteria was inserted into the animals, and their mortality after 48 hours was compared. The study concluded that the propagation of intraabdominal infection and an increase in the severity of disease were "dependent on the composition of the device employed."

It was found that the use of films (Seprafilm and a more pliable Seprafilm II) had a lower mortality rate (37%) as compared to saline control (61%), whereas the gel form of the Seprafilm had a higher mortality rate (85%) than the saline (24%), with statistical significance ($P < .001$). The mortality appeared to be related to the bacteria counts in the blood, but not in the peritoneal fluid. The authors commented that the "increased mortality in gel-treated animals is likely due to a greater number of organisms gaining entry into the bloodstream" ("The increase in bacterial accessibility to the bloodstream following intraperitoneal challenge may reflect the host response to bacterial endotoxin followed by induction of proinflammatory cytokines, vasodilatation, and increased vascular permeability" [Tzianabos et al.[50]]). They also suggested that "*E. coli* is a major contributor to the observed effect." Another interesting finding related to the gel form of the Seprafilm was the effect of heating on the safety of the material. When the gel was exposed to heat at 121°C, it became more transparent. The clearer gel, like saline, resulted in less mortality than the untreated opaque gel. The authors speculated this could be related to the decrease in the molecular weight of the opaque gel (from 85 to 50 kDa) during heating.

The mortality of HA/iron gel at three pH values (4.6, 6.0, and 7.1) was in the range of 90–100%, as compared to the saline group (49%). It was also found that gels containing 4 and 8 mol/L iron potentiated bacterial peritonitis. This is consistent with the knowledge that iron is a factor that may cause marked increase of the virulence of bacterial species.[51,52]

Ruiz-Perez et al.[53] further studied the protection against lethal intraabdominal sepsis (presence of pathogenic organisms or their toxins in the blood or tissues) by HA/CMC-based gels (5%, wt/wt) and found the timing of applying the material a critical factor. HA/CMA gel applied 24 hours before the bacterial challenge did not increase mortality. On the contrary, it protected rats against lethal infection. The study further concluded that HA/CMC/ EDC gels may work as a prodrug[54] when they were hydrolyzed *in vivo*, turning into free HA, CMC and 1-[3-(dimethylamino)propyl]-3-ethylurea (EDU).[53] The immunomodulatory properties of EDU had been tested and confirmed in the rat sepsis model mentioned above, as well as in an endotoxin-induced shock mouse model.[55] No exact evidence, however, was provided with regard to such hydrolysis (of the HA/CMC/EDC gel) in physiological conditions.

5.2.3.4   MeroGel and HylaSine

MeroGel is based on HYAFF, a benzyl ester of HA manufactured by Fidia. HylaSine contains hylan B, a crosslinked HA that is also a component of Synvisc, as mentioned previously (Table 5.4). In the U.S., the regulatory pathways of HA-based products have been vastly different from one category of use (e.g., knee injection or abdominal/OB-GYN surgeries) to another (e.g., nasal and otologic surgery). The former underwent the quite rigorous process of PMA review that required the demonstration of the safety and effectiveness of the products through controlled human clinical trials. The latter, however, was basically making a case that the product is equivalent to another "device," in function and form, already cleared for marketing in the U.S. before 1976 [510(k) is also called Premarket Notification (PMN): "It allows FDA to determine whether the device is equivalent to a device already placed into one of the three classification categories. Thus, 'new' devices (not in commercial distribution prior to May 28, 1976) that have not been classified can be properly identified," http:// www.fda.gov/cdrh/510k.html].

That MeroGel and HylaSine, as HA-based products, were allowed to take the 510(k) regulatory path is based on the legality mentioned above. It is not

TABLE 5.4

HA Products Cleared for Marketing in the U.S. for Use in Nasal and Otologic Surgeries

Product and 510K Number	Product Description	Indications
XOMED MEROGEL Nasal Dressing and Sinus Stent K982731	White fibrous material composed of HYAFF, a derivative of HA through esterification. HA source: rooster combs	For use in nasal/sinus cavities as a dressing or stent, to separate mucosal surfaces, and to help control minimal bleeding
XOMED MEROGEL Otologic Pack K001148	Sheet of non-woven material composed of HYAFF 1 cm × 5 cm or 4 cm × 4 cm HA source: rooster combs	For use in middle ear and external ear canal after canalplasty, myringoplasty, tympanoplasty, and stapes and mastoid surgery, as a dressing or stent, to separate mucosal surfaces, and to help control minimal bleeding
HYLASINE K993362 K012532	Transparent, viscoelastic gel composed of hylan B, a crosslinked HA HA source: rooster combs	For use in nasal/sinus cavities to separate mucosal surfaces, and to help control minimal bleeding following surgery or nasal trauma

exactly clear whether postsurgical application of products containing hylan B to ear and nose (HylaSine) is less risky than injecting it into an arthritic knee joint (Synvisc) or less risky than its use in skin injection, as the case for Hylaform. One may argue, however, that in ear and nose surgery, HA-based material is applied to an open and small cavity and is meant to be present for a limited period of time, whereas the HA materials injected into the joint (Synvisc) or underneath the skin (Hylaform) are intended to last in a closed space inside the body to be effective.

The product safety information published in the 510(k) summary, as in the case of MeroGel and HylaSine, was rather scarce, compared to the information in the PMA summaries of other HA products. Furthermore, there was not much consistency with regard to the amount of information available in different 510(k) summaries. For instance, HylaSine provided data of biocompatibility and pharmacokinetics, as well as human clinical studies. None of such information was available in the 510(k) summary for MeroGel. The predicate products for HylaSine and MeroGel were Merocel (hydroxylated polyvinyl acetal) nasal dressing.

5.2.4 Cosmetic Tissue Augmentation

Crosslinked HA gel can be used as skin filler for cosmetic purposes. It works by adding volume to facial tissue, thus correcting wrinkles and folds and restoring a smooth appearance to the face. To date, two HA-based products, Restylane and Hylaform, have been approved for such indications. Both Restylane and Hylaform are chemically crosslinked HA; the major difference is the chemistry of crosslinking reaction. Restylane uses bis-epoxide crosslinkers, whereas Hylaform uses the same type of crosslinkers for Synvisc (i.e., divinylsulfone; Table 5.5).

A distinct feature of the clinical study design for these two HA-based products is the comparison with other commercially available polymeric material as a positive control. In the studies of HA-based products for ophthalmic, orthopedic, adhesion prevention, the main control groups were either buffered saline or no treatment. For the purpose of wrinkle treatment that is solely dependent on the change of appearance, a positive control, if available, does make much sense. The positive control used in both studies for Restylane and Hylaform was Zyplast, a bovine collagen–based product for wrinkle treatment introduced in 1985 by Inamed Aesthetics (http://www.inamed.com/products), considered the gold standard for such treatment.

5.2.4.1 *Clinical Data in PMAs of Restylane and Hylaform*

The pivotal human clinical study of Restylane submitted to the FDA (P020023) included 128 female and 9 male patients. The adverse events 14 days after the treatment were recorded by patients' diaries, and the follow-up by physicians for 12 months were recorded in case report forms. The adverse events observed at 14 days post–Restylane injection were highlighted in an FDA Talk Paper:

TABLE 5.5

HA Products Cleared for Marketing in the U.S. for the Treatment
of Facial Wrinkles

Product and PMA Number	Product Description	Indications
RESTYLANE Injectable Gel P020023	Viscous, transparent gel composed of NASHA, a crosslinked HA HA source: fermentation Media: PBS (pH 7) [HA] 20 mg/ml Volume: 0.4 ml or 0.7 ml	For mid-to-deep dermal implantation for the correction of moderate to severe facial wrinkles and folds, such as nasolabial folds
HYLAFORM (hylan B gel) P030032	Viscous, clear, colorless gel composed of hylan B, a crosslinked HA HA source: rooster combs Media: 8.5 mg/ml NaCl [HA] 5.5 mg/ml Median particle size: 500 µm	For injection into deep dermis for correction of moderate to severe facial wrinkles and folds, such as nasolabial folds
HYLAFORM PLUS (hylan B gel) P030032/S001	Same as Hylaform except the gel particle size Median particle size: 700 µm	Same as Hylaform

[HA] = concentration of HA or its derivatives, PBS = phosphate buffered solution

"the Restylane treated side had a lower incidence of severe redness (5.1% vs. 5.8%) and an increased incidence of severe bruising (3.6% vs. 0.7%), severe swelling (3.6% vs. 1.4%), severe pain (3.6% vs. 1.4%), and severe tenderness (2.9% vs. 1.4%) compared with the Zyplast treated side" (FDA Talk Paper T03-85, December 12, 2003: "FDA Approves New Product for Facial Wrinkles"). The short-term reactions to Restylane appeared to be more severe than for Zyplast. However, the incidents of such AEs were lower with follow-up injections for both products. It was also important to remember that the design of the clinical study meant that Restylane was injected in one side of the face, and Zyplast in the other side. Therefore, "causality of the systemic adverse events cannot be assigned" (*Summary of Safety and Effectiveness Data, Restylane Injectable Gel, P020023*).

In the study, no hypersensitivity reactions were observed. The hypersensitivity issue of Restylane, however, was not fully resolved because a pre-screening skin test for sensitivity to Restylane was not performed based on the assumption that the hypersensitivity of Restylane should be a nonissue ("Pre-screening skin testing for sensitivity to the crosslinked collagen Control was performed. Pre-screening skin test for sensitivity to Restylane was not performed due to low suspicion of hypersensitivity," *Summary of Safety and Effectiveness Data, Restylane Injectable Gel, P020023*). A more scientific approach to this issue was to evaluate antibody titers before and after treatment with Restylane. The FDA stated that "the overlap of symptom profiles for Restylane hypersensitivity and injection site reactions, and lack of correlation of symptoms with anti-body titers, may have confounded diagnosis of hypersensitivity reaction" to Restylane. Therefore, CDRH required a follow-up, postmarketing study "to assess the likelihood of hypersensitivity reactions due to injection of Restylane."

In the pivotal clinical trial to evaluate the safety (and effectiveness) of Hylaform (P030032), 261 patients were randomly assigned between the Hylaform (133 subjects) and Zyplast control (128 subjects) at 10 dermatology centers in the U.S. Patients were followed for 12 weeks after receiving injection of the dermal filler in the nasolabial folds.

It was concluded that in the Hylaform studies, the device-related adverse events did not occur frequently in either group and were primarily injection site erythema, bruising, swelling, pain, nodules, pruritus, and tenderness of "mild intensity." No information 2 weeks postinjection was available, as was the case for Restylane. Hylan B IgG antibody titers were measured at baseline and throughout the treatment. Among the 133 patients who received Hylaform, only one patient exhibited a positive antibody response (*Summary of Safety and Effectiveness, Hylaform, P030032*).

In both the cases for Restylane and Hylaform, the "reasonable assurance" of safety appeared to be established "by the lack of severe adverse events, and by the short duration of the events observed." (Restylane, P020023). Before the development of HA-based soft-tissue augmentation products, however, there were already on the market a few dozen products with the same or similar indications. The main scientific rationale for developing HA-based products was the anticipation that HA-based products should be not just as safe but much safer than the collagen-based products. In fact, the safety profiles of Restylane and Hylaform have turned out to be less satisfactory than originally expected.

However, this should not come as a surprise, because chemical crosslinking altered HA size and its structure as well. The biocompatibility of chemically crosslinked HA needs to be examined on the basis of the experimental evidence of each individual product. Many factors may affect the biological nature of the chemically altered HA products, including the type of the chemistry of crosslinking and the formulation.

5.2.4.2   *Adverse Events of Restylane and Hylaform in Medical Journals*

The pivotal study of Restylane submitted to the FDA for approval, as mentioned earlier, was later published in a medical journal.[56] There have been also some reports since 2000 by individual clinicians with regard to the experience of adverse events of Restylane and Hylaform. Lupton and Alster reported the first case of a patient developing an acute hypersensitivity reaction (an adverse granulomatous-like response) to locally injected hyaluronan gel from a nonanimal source (Restylane).[57] An adult female received three rounds of intradermal injections approximately half a year apart. The first two were uneventful, but 2 weeks after the third series of injections, she developed acute multiple, tender nodules with a size of up to 1 cm diameter. Some exuded puslike material, and others were more indurated (hardened) and almost fibrotic. The bacteria culture of the drain from the cystic nodules was negative. Lupton et al. believed that "despite previous reports to the contrary, injectable hyaluronic acid gel may be capable of producing an allergic-type reaction." The authors recommended that patients be warned of this potential complication and pointed out that as (in this case) the symptoms developed only after the third treatment, even skin test for allergy would not have elicited the allergic reaction.

Raulin et al. reported an exudative granulomatous foreign body reaction for Hylaform[58] similar to those found with collagen-based products such as Zyderm and Zyplast. The authors, however, could not determine whether the process was immunological or allergic. The reproducibility of the reaction in skin test points to an allergic reaction, whereas the formation of granulomas indicates immunological pathogenesis. The complete regression of the granulomas in this case occurred within 6 weeks, whereas the effect of such reactions with collagen normally lasts 4–6 months, according to the study by Castrow and Krull.[59]

Granulomatous foreign body reactions against bovine collagen were reported to have an incidence rate of 1.3%.[59] Although there are no data to indicate that crosslinked HA-based products also have a similar rate of incidence, these cases indicate that "HA does not seem to be devoid of this risk."

Fernández-Aceñero et al. in Madrid, Spain,[60] reported a case of a female patient who developed nodules in the lips after Restylane injection for lip augmentation. The histological analysis revealed the presence of a granulomatous reaction surrounding a blue amorphous material. The material was confirmed to contain HA, as it stained intensely with Alcian Blue. The reaction for the patient occurred after the first injection of Restylane. The duration of the reaction was unknown, and its cause and source was not exactly clear. The authors postulated that because their patient had never received a HA injection before, and the reaction appeared after the first injection, the adverse reaction could be a result of potential impurities in the material injected.

Micheels, in Geneva, Switzerland,[61] treated patients with crosslinked HA (133 with Restylane, 106 with Hylaform) from 1997 to 2001 and observed eight cases of giant cell reactions that lasted up to 4.5 months. The clinical symptoms were redness, pruritus, painful swelling of the injected area, and a nettle-type of rash reaction in the treated area in two patients. The ratio of adverse reactions over the total population is greater than 1 in 30. Kavouni and Stanec of London reported a similar case[62] in response to the paper by Micheel: "Three days after the Restylane injections the patient presented with erythematous swelling of the treated area and mildly tender nodules at the injection sites. . . . The clinical manifestations of this reaction are consistent with a delayed hypersensitivity type of inflammatory reaction." On this occasion, fortunately, the reaction was self-limiting, and no residual intradermal lesions could be identified at follow up. Kavouni also quoted a personal communication from June 2001 with the manufacturer of Restylane — Q-Med — who asserted that this type of reaction is rare, having an incidence of only 1 in 5000. Obviously, there is discrepancy between the rates of such an adverse reaction reported from different sources.

Some studies of Restylane, especially those closely associated with the manufacturers of HA-based dermal fillers, concluded that the adverse reactions of Restylane were not a result of the intrinsic nature of the product but, rather, the impurities in the HA raw material, and that the problem had been identified and resolved.[63] Friedman et al. (the leading author of this study, Dr. Geronemus, was a member of the Scientific Advisory Board of Q-Med) looked retrospectively at the data on the safety of NASHA Gel (Restylane, Perlane, Restylane Fine Lines) in Europe, Canada, Australia, South America, and Asia from 1999 and 2000,[63] and in particular at the incidence of localized hypersensitivity reactions. In 1999, an estimated 144,000 patients were treated with this type of crosslinked HA gel for soft-tissue augmentation, and approximately 1 in every 1400 patients developed localized hypersensitivity reactions. In 2000, in the estimated 262,000 treated patients, only 1 in 5000 patients had such adverse reactions. The paper attributed the declining of incidence of hypersensitivity from 1999 to 2000 to the introduction of a more purified HA raw material. In mid-1999, Q-Med introduced HA raw material with amount of proteins reduced to the range of 13–17 µg/ml of product; that is, six times lower than the raw material previously used and approximately the same as that found in Hylaform. The review also pointed out that the higher incidence reported by Micheels[61] (1/30) and Lowe[64] (3/709 or 0.42%) was mostly from the batches manufactured before 2000.

What may actually elicit antibody? The often-mentioned theories are protein impurity in HA, contaminant or residue chemicals (diepoxide), or toxic degradation products of crosslinked HA. Another distinct possibility is that the crosslinked HA per se sometimes could be immunogenic, as Micheels surmised, "if hyaluronic acid does not show species and organ specificity, could its stereometric manipulation through crosslinkage/stabilization cause it to be acknowledged as a foreign body by an organism and induce antibody production?" This possibility apparently cannot be ruled out. As we know, the variability does exist among batches in the manufacturing of HA, especially crosslinked HA. This could explain why the same brand of product may not elicit the same type or degree of adverse reactions all the time.

In fact, although serum analysis revealed positive antibodies against Restylane and Hylaform, patients having reactions to extracted animal HA showed negative antichicken protein antibodies. This seems to indicate that animal origin of HA is not the cause of the problem.[61] Moreover, the purification processes of HA from rooster combs have been tested and industrialized for decades. The HA products based on animal sources, such as Healon, Amvisc, and Hyalgan, have proved to be as safe as HA from nonanimal sources.

Although the safety of HA-based products for cosmetic interdermal injection is generally acceptable, this type of product is not risk free. In addition to the potential risk of injury related to the injection procedure, the property of individual products may have a certain role. The cause can be impurity or crosslinking itself, as determined by the type of chemistry, batch to batch variation, and so on. Therefore, patients should be informed of the possibility of such complications, and similar to collagen-based products, the skin test for HA-based gel may be justified as one of such precautionary measures.

References

1. Balazs, E. A., in *Healon (Sodium Hyaluronate), A Guide to Its Use in Ophthalmic Surgery*, Miller, D. and Stegmann, R., Eds., Wiley, New York, 1983, pp. 5–27.
2. Diamond, M. P., Reduction of adhesions after uterine myomectomy by Seprafilm membrane (HAL-F): a blinded, prospective, randomized, multicenter clinical study. Seprafilm Adhesion Study Group, *Fertil. Steril.*, 66, 904–910 (1996).
3. Johns, D. B., Keyport, G. M., Hoehler, F., and diZerega, G. S., Reduction of postsurgical adhesions with Intergel adhesion prevention solution: a multi-center study of safety and efficacy after conservative gynecologic surgery, *Fertil. Steril.*, 76, 595–604 (2001).
4. Lussier, A., Cividino, A. A., McFarlane, C. A., Olszynski, W. P., Potashner, W. J., and De Medicis, R., Viscosupplementation with hylan for the treatment of osteoarthritis: findings from clinical practice in Canada, *J. Rheumatol.*, 23, 1579–1585 (1996).
5. Maheu, E., Ayral, X., and Dougados, M., A hyaluronan preparation (500-730 kDa) in the treatment of osteoarthritis: a review of clinical trials with Hyalgan, *Int. J. Clin. Pract.*, 56, 804–813 (2002).
6. Lohmander, L. S., Dalen, N., Englund, G., Hamalainen, M., Jensen, E. M., Karlsson, K., Odensten, M., Ryd, L., Sernbo, I., Suomalainen, O., and Tegnander, A., Intra-articular hyaluronan injections in the treatment of osteoarthritis of the knee: a randomized, double blind, placebo controlled multicentre trial, *Ann. Rheum. Dis.*, 55, 424–431 (1996).
7. Namiki, O., Toyoshima, H., and Morisaki, N., Therapeutic effect of intra-articular injection of high molecular weight hyaluronic acid on osteoarthritis of the knee, *Int. J. Clin. Pharmacol. Ther. Toxicol.*, 20, 501–507 (1982).
8. Reed, R. K. and Laurent, U. B., Turnover of hyaluronan in the microcirculation, *Am. Rev. Respir. Dis.*, 146, S37–S39 (1992).
9. Reed, R. K., Lilja, K., and Laurent, T. C., Hyaluronan in the rat with special reference to the skin, *Acta Physiol. Scand.*, 134, 405–411 (1988).
10. Laurent, T. C., in *Chemistry and Molecular Biology of the Intercellular Matrix*, Balazs, E. A., Ed., Academic Press, London, 1970, pp. 703–732.
11. Laurent, T. C., Dahl, I. M., Dahl, L. B., Engstrom-Laurent, A., Eriksson, S., Fraser, J. R., Granath, K. A., Laurent, C., Laurent, U. B., Lilja, K., The catabolic fate of hyaluronic acid, *Connect. Tissue Res.*, 15, 33–41 (1986).
12. Fraser, J. R., Kimpton, W. G., Laurent, T. C., Cahill, R. N., and Vakakis, N., Uptake and degradation of hyaluronan in lymphatic tissue, *Biochem. J.*, 256, 153–158 (1988).
13. Smedsrod, B., Cellular events in the uptake and degradation of hyaluronan, *Adv. Drug Delivery Rev.*, 7, 265–278 (1991).
14. Laurent, U. B., Dahl, L. B., and Reed, R. K., Catabolism of hyaluronan in rabbit skin takes place locally, in lymph nodes and liver, *Exp. Physiol.*, 76, 695–703 (1991).

15. Tammi, R. H., Tammi, M. I., Hascall, V. C., Hogg, M., Pasonen, S., and MacCallum, D. K., A preformed basal lamina alters the metabolism and distribution of hyaluronan in epidermal keratinocyte "organotypic" cultures grown on collagen matrices, *Histochem. Cell Biol.*, 113, 265–277 (2000).

16. Tammi, R., MacCallum, D., Hascall, V. C., Pienimaki, J. P., Hyttinen, M., and Tammi, M., Hyaluronan bound to CD44 on keratinocytes is displaced by hyaluronan decasaccharides and not hexasaccharides, *J. Biol. Chem.*, 273, 28878–28888 (1998).

17. Tammi, R., Saamanen, A. M., Maibach, H. I., and Tammi, M., Degradation of newly synthesized high molecular mass hyaluronan in the epidermal and dermal compartments of human skin in organ culture, *J. Invest. Dermatol.*, 97, 126–130 (1991).

18. Tammi, R. H., Pasonen-Seppanen, S., Kultti, A., Hyttinen, J. M. T., MacCallum, D. K., Hascall, V. C., and Tammi, M. I., Hyaluronan degradation in epidermis, Poster 30, *Hyaluronan 2003 Conference*, Cleveland, OH, 2003.

19. Stair-Nawy, S., Csoka, A. B., and Stern, R., Hyaluronidase expression in human skin fibroblasts, *Biochem. Biophys. Res. Commun.*, 266, 268–273 (1999).

20. Reed, R. K., Laurent, U. B., Fraser, J. R., and Laurent, T. C., Removal rate of [3H]hyaluronan injected subcutaneously in rabbits, *Am. J. Physiol.*, 259, H532–H535 (1990).

21. Fraser, J. R. and Laurent, T. C., Turnover and metabolism of hyaluronan, *Ciba Found. Symp.*, 143, 41–45 (1989).

22. Arbogast, B., Hopwood, J. J., and Dorfman, A., Absence of hyaluronidase in cultured human skin fibroblasts, *Biochem. Biophys. Res. Commun.*, 67, 376–382 (1975).

23. Roden, L., Campbell, P., Fraser, J. R., Laurent, T. C., Pertoft, H., and Thompson, J. N., Enzymic pathways of hyaluronan catabolism, *Ciba Found. Symp.*, 143, 60–67 (1989).

24. Fraser, J. R., Laurent, T. C., Engstrom-Laurent, A., and Laurent, U. G., Elimination of hyaluronic acid from the blood stream in the human, *Clin. Exp. Pharmacol. Physiol.*, 11, 17–25 (1984).

25. Floren, I., Viscoelastic purity, *J. Cataract Refract. Surg.*, 24, 145–146 (1998).

26. Dick, H. B., Augustin, A. J., Pakula, T., and Pfeiffer, N., Endotoxins in ophthalmic viscosurgical devices, *Eur. J. Ophthalmol.*, 13, 176–184 (2003).

27. Folk, J. C., Packer, A. J., Weingeist, T. A., and Howcroft, M. J., Sodium hyaluronate (Healon) in closed vitrectomy, *Ophthal. Surg.*, 17, 299–306 (1986).

28. Vatne, H. O. and Syrdalen, P., The use of sodium hyaluronate (Healon) in the treatment of complicated cases of retinal detachment, *Acta Ophthalmol. (Copenh.)*, 64, 169–172 (1986).

29. Jones, A., Regan, M., Ledingham, J., Pattrick, M., Manhire, A., and Doherty, M., Importance of placement of intra-articular steroid injections, *Br. Med. J.*, 307, 1329–1330 (1993).

30. Goldberg, D. L., Brandt, K. D., and Cohen, A. S., Rapid, simple detection of trace amounts of synovial fluid, *Arthritis Rheum.*, 16, 487–490 (1973).

31. Hamburger, M. I., Lakhanpal, S., Mooar, P. A., and Oster, D., Intra-articular hyaluronans: a review of product-specific safety profiles, *Semin. Arthritis Rheum.*, 32, 296–309 (2003).

32. Bucher, W., Otto, T., and Hamburger, M. I., Differentiation of hyaluronate products by qualitative differences in their immunogenicity in rabbits: possible mechanism for product-specific severe adverse reactions? Comment on the article by Martens, *Arthritis Rheum.*, 46, 2538–2548 (2002).

33. Ali, Y., Weinstein, M., and Jokl, P., Acute pseudogout following intra-articular injection of high molecular weight hyaluronic acid, *Am. J. Med.*, 107, 641–642 (1999).

34. Yacyshyn, E. A. and Matteson, E. L., Gout after intra-articular injection of Hylan GF-20 (Synvisc), *J. Rheumatol.*, 26, 2717 (1999).

35. Kroesen, S., Schmid, W., and Theiler, R., Induction of an acute attack of calcium pyrophosphate dihydrate arthritis by intra-articular injection of Hylan G-F 20 (Synvisc), *Clin. Rheumatol.*, 19, 147–149 (2000).

36. Luzar, M. J. and Altawil, B., Pseudogout following intraarticular injection of sodium hyaluronate, *Arthritis Rheum.*, 41, 939–940 (1998).

37. Maillefert J. F., Hirschhorn, P., Pascaud F., Piroth C., and Tavernier C., Acute attack of chondrocalcinosis after an intraarticular injection of hyaluronan, *Rev. Rhum. Engl. Ed.*, 64, 593–594 (1997).

38. Puttick, M. P. E., Wade, J. P., Chalmers, A., Connell, D. G., and Rangno, K. K., Acute local reactions after intra-articular Hylan for osteoarthritis of the knee, *J. Rheumatol.*, 22, 1311–1314 (1995).

39. Allen, E. and Krohn, K., Adverse reaction to Hylan GF-20, *J. Rheumatol.*, 27, 1572 (2000).

40. Martens, P. B., Bilateral symmetric inflammatory reaction to Hylan G-F 20 injection, *Arthritis Rheum.*, 44, 978–979 (2001).

41. Rees, J. D. and Wojtulewski, J. A., Systemic reaction to viscosupplementation for knee osteoarthritis, *Rheumatology*, 40, 1425–1426 (2001).

42. Leopold, S. S., Warme, W. J., Pettis, P. D., and Shott, S., Increased frequency of acute local reaction to intra-articular Hylan GF-20 (Synvisc) in patients receiving more than one course of treatment, *J. Bone Joint Surg.*, 84A, 1619–1623 (2002).

43. Pullman-Mooar, S., Mooar, P., Sieck, M., Clayburne, G., and Schumacher, H. R., Are there distinctive inflammatory flares after Hylan G-F 20 intraarticular injections?, *J. Rheumatol.*, 29, 2611–2614 (2002).

44. Zardawi, I. M. and Chan, I., Synvisc perisynovitis, *Pathology*, 33, 519–520 (2001).

45. Chen, A. L., Desai, P., Adler, E. M., and DiCesare, P. E., Granulomatous inflammation after Hylan G-F 20 viscosupplementation of the knee: a report of six cases, *J. Bone Joint Surg.*, 84A, 1142–1147 (2002).

46. Mont, M. and Etienne, G., Sequelae of Hylan G-F 20 viscosupplementation of the knee, *J. Bone Joint Surg. Am.*, 85A, 967–968 (2003).

47. Morton, A. and Shannon, P., Increased frequency of acute local reaction to intra-articular Hylan G-F 20 (Synvisc) in patients receiving more than one course of treatment, *J. Bone Joint Surg.*, 85-A, 2050 (2003).

48. Sasaki, M., Miyazaki, Y., and Takahashi, T., Hylan g-f 20 induces delayed foreign body inflammation in guinea pigs and rabbits, *Toxicol. Pathol.*, 31, 321–325 (2003).

49. Beck, D. E., Cohen, Z., Fleshman, J. W., Kaufman, H. S., van Goor, H., and Wolff, B. G., A prospective, randomized, multicenter, controlled study of the safety of Seprafilm adhesion barrier in abdominopelvic surgery of the intestine, *Dis. Colon Rectum.*, 46, 1310–1319 (2003).

50. Tzianabos, A. O., Cisneros, R. L., Gershkovich, J., Johnson, J., Miller, R. J., Burns, J. W., and Onderdonk, A. B., Effect of surgical adhesion reduction devices on the propagation of experimental intra-abdominal infection, *Arch. Surg.*, 134, 1254–1259 (1999).

51. Neilands, J. B., Siderophores: structure and function of microbial iron transport compounds, *J. Biol. Chem.*, 270, 26723–26726 (1995).

52. Biosca, E. G., Fouz, B., Alcaide, E., and Amaro, C., Siderophore-mediated iron acquisition mechanisms in *Vibrio vulnificus* biotype 2, *Appl. Environ. Microbiol.*, 62, 928–935 (1996).

53. Ruiz-Perez, B., Cisneros, R. L., Matsumoto, T., Miller, R. J., Vasios, G., Calias, P., and Onderdonk, A. B., Protection against lethal intra-abdominal sepsis by 1-(3-dimethylaminopropyl)-3-ethylurea, *J. Infect. Dis.*, 188, 378–387 (2003).

54. Ruiz-Perez, B., Cisneros, R. L., Matsumoto, T., Miller, R. J., Vasios, G., Calias, P., Onderdonk, A. B., Tzianabos, A. O., Gershkovich, J., Johnson, J., Burns, J. W., and Stavesk, M., Therapeutic utility of chemically modified hyaluronan, Poster 62, *Hyaluronan 2003 Conference,* Cleveland, OH (2003).

55. Matsumoto, T., Nieuwenhuis, E. E., Cisneros, R. L., Ruiz-Perez, B., Yamaguchi, K., Blumberg, R. S., and Onderdonk, A. B., Protective effect of ethyl-3-(3-dimethyl aminopropyl)urea dihydrochloride (EDU) against LPS-induced death in mice, *J. Med. Microbiol.*, 53, 97–102 (2004).

56. Narins, R. S., Brandt, F., Leyden, J., Lorenc, Z. P., Rubin, M., and Smith, S. A., A randomized, double-blind, multicenter comparison of the efficacy and tolerability of Restylane versus Zyplast for the correction of nasolabial folds, *Dermatol. Surg.*, 29, 588–595 (2003).

57. Lupton, J. R. and Alster, T. S., Cutaneous hypersensitivity reaction to injectable hyaluronic acid gel, *Dermatol. Surg.*, 26, 135–137 (2000).

58. Raulin, C., Greve, B., Hartschuh, W., and Soegding, K., Exudative granulomatous reaction to hyaluronic acid (Hylaform), *Contact Dermatitis*, 43, 178–179 (2000).

59. Castrow, F. F. N. and Krull, E. A., Injectable collagen implant — update, *J. Am. Acad. Dermatol.*, 9, 889–893 (1983).

60. Fernández-Aceñero, M. J., Zamora, E., and Borbujo, J., Granulomatous foreign body reaction against hyaluronic acid: report of a case after lip augmentation, *Dermatol. Surg.*, 29, 1225–1226 (2003).

61. Micheels, P., Human anti-hyaluronic acid antibodies: is it possible?, *Dermatol. Surg.*, 27, 185–191 (2001).

62. Kavouni, A. and Stanec, J. J., Human antihyaluronic acid antibodies, *Dermatol. Surg.*, 28, 359–360 (2002).

63. Friedman, P. M., Mafong, E. A., Kauvar, A. N. B., and Geronemus, R. G., Safety data of injectable nonanimal stabilized hyaluronic acid gel for soft tissue augmentation, *Dermatol. Surg.*, 28, 491–494 (2002).

64. Lowe, N. J., Maxwell, C. A., Lowe, P., Duick, M. G., and Shah, K., Hyaluronic acid skin fillers: adverse reactions and skin testing, *J. Am. Acad. Dermatol.*, 45, 930–933 (2001).

6

Clinical Performance, Mechanism of Action, and Product Characteristics

Chapter 1 listed major hyaluronan (HA)–based products approved so far by the U.S. Food and Drug Administration (FDA; Table 1.1). Chapter 5 provided more detailed information with regard to their compositions, sources, formulae, and indications (Tables 5.1–5.5). Safety and effectiveness are the two essential requirements for a medical device to gain the regulatory approval from the FDA. It is understood that all HA products that passed the premarket scrutiny of the FDA were deemed safe and effective for their indicated use at the time of approval. The challenge to the products, however, continued after they officially entered the market. The extensive application of a product in a real, broad market place is the ultimate validation of the usefulness of the product. The customers as well as clinicians, engineers, and scientists continue to examine to what extent the HA products actually meet the rationale and expectation of their original design. How effective is an HA-based product in comparison with its alternatives for its intended use? How beneficial is that product, in balance with its potential risk and economic cost? How sound is the scientific rationale behind the product design and development?

These questions are better answered after years of postmarketing clinical experience. As extensive technological, clinical, and regulatory information regarding HA-based products becomes more available, so is it more feasible for a more scientific and objective assessment of the overall merits of those products. This would help to improve the products and to enhance the public's awareness of the products' benefits and risks, as well.

The opinions of clinicians with experience with HA-based products can be especially valuable. Certain insight can only come from those who use the product to tackle real medical problems in the real world. The feedback from the end users, clinicians and patients alike, provides clues to what kind of properties of the HA-based products are truly relevant and important to their clinical performance, and to what extent a certain product has succeeded or failed to function as intended. Analyzing postmarketing clinical information is a means to evaluate the product design and, in that sense, serves as an extension of product development.

6.1   Ophthalmic Surgeries

Viscoelasticity and transparency are the main required characteristics of those HA-based products for tissue protection during ophthalmic surgeries. In cataract extraction surgery, viscosity aids in maintaining a deep anterior chamber. In retinal attachment surgery, viscoelastic HA material serves as a vitreous replacement and aids in holding the retina against the sclera for reattachment. In ophthalmic surgical procedures, HA viscoelasticity has a "soft spatula" effect that softens the impact of surgical instruments on the extremely delicate eye tissues. The importance of transparency is obvious, in that a material used as a surgical aid should not obscure the vision of the surgeons. The clearness of natural HA has rarely been a problem, unless impurities, including particulates, are not adequately removed. Because HA for ophthalmic use has always been polymeric HA with a molecular weight (MW) above certain level (mostly above 1 million Da), and at a normal range of concentration (around or above 1% w/v), the viscous property of the products becomes obvious by the naked eye. Such was the rationale for the original product design.

Over the last couple of decades, HA-containing products have gained wide acceptance and even become a preferred choice for many ophthalmic applications especially in anterior segment surgeries. The extensive feedback from surgeons and independent research has furthered the understanding of the mechanism of HA action and helped extend the HA product lines tailored for different kinds of eye surgeries. The progress in product development has gone hand in hand with the progress in the quantitative determination of the effect of product parameters on product performance. From a material science viewpoint, it is always worthy of continued inquisition into the interrelationship between the product features and their functions. The following are a few interesting examples.

6.1.1   Cataract Removal — Intraocular Lens Implantation

6.1.1.1   Viscoelasticity and Lubricity

A certain degree of viscosity is required for HA to remain in the anterior chamber to prevent tissue–tissue contact. Lang et al. conducted studies[1] to determine the level of viscosity of a material considered useful as an aid in anterior segment surgery. They found that Healon (NaHA), with a concentration of 10 mg/ml and a viscosity of 270 poise, is gel-like and remains in the eye even when external pressure is applied to the cornea. NaHA at 83–150 poise (8–9 mg/ml) creeps out of the eye under external pressure or corneal retraction. NaHA at 17 poise (6 mg/ml) flows out of the eye when the cornea is retracted, similar to the chondroitin 6-sulfate with a concentration of 500 mg/ml (13.5 poise). The authors concluded that a threshold of 80 poise was needed for useful performance in the surgery. All measurements were done at shear rate of about 10 second^{-1}, a close approximation of the shear rate under many surgical conditions.

As noted in Chapter 3, the value of viscosity of HA can be correctly interpreted only within the context of the shear rate at which the viscosity is measured. The HA viscoelastic material with a viscosity of 270 poise at a shear rate of 10 second^{-1} actually displays much higher viscosity at rest. According to Bothner and Wik,[2] the zero shear rate viscosity of a 10 mg/ml NaHA with a MW of approximately 3 million dalton can reach 1000 poise. It is the high viscosity that prevents the egress of HA from the surgically opened eye under compressive forces (nonshear forces) of the cornea and the iris-lens-vitreous diaphragm. However, the viscosity of the same HA material can be reduced to 1 poise when it is subjected to a high shear rate of 1000 second^{-1}, such as in the situation when the viscoelastic material passes through a thin cannula (e.g., 27 gauge). This shear-thinning phenomenon is also known as pseudoplasticity.

Pseudoplasticity of a viscous material can be quantified as the quotient of two different viscosities at two different shear rates. Customarily, it is often given as the quotient of the viscosity at rest over the viscosity at a shear rate of 100/second. Dick et al. reported[3] the mean pseudoplasticity of Healon (173), Healon GV (754), and Healon5 (591). Be aware that pseudoplasticity merely reflects the relative effect of the shear force on the viscosity of the material. Pseudoplasticity by itself does not determine the absolute value of viscosity under different conditions, and a higher pseudoplasticity does not necessarily mean a lower viscosity at the same high shear rate. For example, Healon5 had a zero shear viscosity of 5525 Pas, with a pseudoplasticity number of 591. At a shear rate of 100/second, the viscosity of Healon5 would be 9.36 Pas (5525/591). Likewise, Healon has a viscosity of 243 Pas at rest. With a pseudoplasticity number of 173, its viscosity at a shear rate of 100/second would be 1.40 Pas (243/173). Even with a much higher pseudoplasticity, Healon5 is still more viscous than Healon at a shear rate of 100/second.

When properly formulated within a certain range of MW and concentration, HA is the best biomaterial known to be both viscous and lubricious. In addition to making it easier to expel HA through a thin cannula, shear-thinning has important clinical ramifications. Whereas viscosity holds the space between delicate tissues and elasticity softens the effect of surgical instruments, pseudoplasticity facilitates the maneuvering of the surgical instruments and handling of the material. Conceivably, pseudoplasticity may enable the HA layer not in immediate contact with the tissue surface to move under shear force with less resistance than Newtonian fluid (chondroitin sulfate) of the same viscosity at rest. This reduces the abrasion such movement might impose on the tissue that the viscoelastic material is supposed to protect.

6.1.1.2 *Viscosity and Thickness of Coating*

Even with its shear-thinning property, a thin layer of highly viscous HA material could be problematic to endothelial cells under certain circumstances, such as during the insertion of intraocular lens, as demonstrated by the study of Hammer and Burch.[4] In that study, bovine corneal buttons were covered with a thin layer (~0.5 mm) of HA, chondroitin sulfate (CS), methylcellulose, or saline and abraded twice with an intraocular lens (IOL) at a speed of 2–3 cm/second (shear rate 40–60 second^{-1}). Then the damage to endothelial cells was evaluated under a microscope.

The high-MW HA used in the study was Healon, a 1% NaHA solution, and the HA samples with lower concentrations were prepared by diluting Healon with a salt solution. Three of the tested materials had a medium range of viscosity (30 cP), and they were prepared with 0.17% high-MW HA (sample 1); 20% CS (sample 2), and 1% methylcellulose (sample 3), respectively. Sample 4 was a balanced salt solution (0.70 cP), sample 5 was a thin layer of 1% high-MW HA and sample 6 was a thick layer of 1% high-MW HA. Interestingly, samples 1, 2, 3, and 6 showed less than 5% cell damage, whereas the damage from sample 5 was nearly 30%, and that from sample 4 (saline) was nearly 50%.

What is the explanation? First, the thickness of the layer of viscoelastic has much to do with the force on the cells. It is consistent with the definition of dynamic viscosity: η = shear force/shear rate.

Shear force is f/A, and shear rate is dv/dy (where f is force, A is area, v is velocity, and y is the distance perpendicular to the applied force). Therefore, η can be rewritten as

$$\eta = fdy/Adv$$

It is concluded from the equation that for a certain material with a certain viscosity (η), the force f (the damage to cells) is inversely proportional to the distance y (the thickness). Conversely, when y is fixed, f is directly proportional to η, the viscosity.

Hammer and Burch theorized that when an intraocular lens is drawn past the corneal endothelium at 2–3 cm/second covered with a thin layer (0.5 mm coating) of a protective agent, the shear rate is approximately 40–60 second^{-1}. At that shear rate and a medium viscosity of 30 cP, the shear force on the corneal endothelium was 15 dynes/cm^2, whereas the shear force for a thin layer of 1% NaHA reached 500 dynes/cm^2. This excessive shear force had a deleterious effect on the endothelium. At the other end of the spectrum, the balanced salt solution, naturally with low viscosity, failed to provide meaningful protection against intraocular lens abrasion as well. It appears from Hammer's study that the protection of the cornea requires a good coating material with medium viscosity.

Glasser et al. conducted a similar intraocular-lens abrasion test[5] using rabbit cornea for an *in vitro* comparison of endothelial protection offered by four viscous solutions of Healon (1% NaHA), Amvisc (1% NaHA), Viscoat (4% CS and 3% HA), and 2% methylcellulose. All four solutions provided complete endothelial protection from mechanical trauma. Endothelial cell density and morphology were unaffected in an *in vivo* toxicity study of cat cornea. Note that the 0.2 ml was not a small amount to be placed on the surface of a rabbit corneal button, as compared with 0.03 ml placed on a bovine cornea in Hammer's study discussed earlier. Therefore, the layer created in Glasser's study was probably much thicker than 0.5 mm. This could be the reason why no difference of protection by the four materials was found in the study by Glasser, as one might have anticipated based on the study conducted by Hammer.

Glasser et al. also compared the protection to rabbit corneal endothelium offered by Healon, Viscoat, and BSS Plus (irrigation solution) during pha-coemulsification with and without traumatic intraocular lens implantation.[6] There was no significant difference in cell damage between groups during phacoemulsification alone. Cell damage after traumatic lens insertion, how-ever, was significantly greater in the groups treated with BSS Plus (76.2%) and Healon (41.4%) than in either paired Viscoat-treated group (21.1% and 17.4%).

Viscoat was noted to be adherent to the cornea at the end of the procedure in one-third of the cases, indicating that significant amount of Viscoat remained in the anterior chamber during surgery. Conversely, Healon did not adhere to the cornea during the procedures. Glasser appeared to agree with the conclusion drawn by Hammer that as Healon flowed out of the chamber during phacoemulsification, aspiration, and irrigation, only a thin layer of highly viscous Healon remained, which "allows high shear forces to be transmitted to the endothelium, resulting in significant damage."

McDermott et al. conducted a rabbit study to quantify the viscoelastic adher-ence to the corneal endothelium (i.e., the thickness of the layer of viscoelastics left after phacoemulsification).[7] They found that the "median viscoelastic thicknesses were 13.0 μm for Amvisc Plus, 0.4 μm for Healon GV, and 375.0 μm for Viscoat. Each was significantly different from the others (Kruskal-Wallis, $P < .001$)." This appears to support the notion that a thin layer of highly viscous material may be responsible for extra endothelial cell loss.

6.1.1.3 Intraocular Pressure, Osmosis, and Corneal Edema

One of the safety concerns associated with the application of HA and other viscoelastics in eye surgery is the postsurgical increase of intraocular pres-sure. It is widely believed that an ophthalmic viscosurgical device (OVD) product with high MW and high viscosity is more likely to cause postop-erative intraocular pressure (IOP) increase. Conceivably, it is more difficult for larger molecules, with high viscosity, to pass through the trabecular meshwork — a network of fibers responsible for draining the aqueous humor from the eye. The experimental data, however, disagree. Following are two examples.

Schubert et al.[8] compared the intraocular pressures of owl monkeys given 1% NaHA solutions with a range of different MWs (1.7, 3.4, 3.7, 4.5, and 4.9×10^6) and found that "the highest and most persistent increase in IOP was observed after the injection of the solution with the lowest viscosity (10,000 cSt)." The smallest postoperative IOP increase was observed after replacing aqueous humor with those HA samples with viscosities of 100,000 (MW 3.7×10^6) to

300,000 cSt (MW 4.5×10^6). At the same time, it was found that the highly viscous, high-MW NaHA (930,000 cSt), although having a longer half-life, could still diffuse through the channels of the anterior chamber, and its concentration was reduced to approximately one-third after 24 hours. Concurrently, it was observed that the very high value of the IOP was associated with the very low value of the half-life of the material.

Fry compared the effect of four OVDs on the IOP increase at 4 and 8 hours after surgery.[9] The low-MW (HA and CS) and high-concentration Viscoat, when left without being aspirated after surgery, had the highest intraocular pressure increase, followed in decreasing order by the Viscoat group (with aspiration), the Amvisc group (with aspiration), and the Healon group (with aspiration). Fry concluded that Viscoat should be aspirated after surgery, in spite of the opposite claim by the manufacturer at the time.

The increase of intraocular pressure in the eye is caused by the net influx of the liquid, which in large part is driven by osmotic pressure. As a colligative property, osmotic pressure mainly depends on the number of molecules (molar concentrations), not on the nature of the molecules (whether NaHA, CS, or NaCl). Osmosis is the movement of water across a semipermeable membrane from a region of lower solute concentration to one of higher solute concentration. When a significant amount of exogenous HA residue is present in a more or less sealed compartment like an eye, the hypertonic solution inside the eye causes osmosis from outside, which leads to the increase of IOP. Other factors being equal, such as the effect of electrolytes and other small molecules, a polymer solution with higher molar concentration should have higher colloidal osmotic pressure than that of the less concentrated formulation.

Viscoat (30 mg of HA and 40 mg of CS) has seven times more polymeric material than that of Healon (10 mg/ml of HA). The ratio of number of molecules of Viscoat versus Healon is even higher, considering the lower MW of HA and CS (estimated MW 50,000, per Merck Index 2270) in Viscoat as compared to Healon. That contributes to the higher IOP of Viscoat postoperatively. Note that the degree and direction of postsurgical IOP change also depends on when it is measured. It is conceivable that high–molar concentration preparations of small molecules may cause a sharp increase of IOP, which subsides rapidly as the preparation exits the eye, facilitated by the initial burst of IOP. This appeared to be the case in the study of rabbit eye by Mac Rae et al.[10] A 20% CS solution was prepared with an osmolality of 656 mOsm, approximately twice as high as most isotonic OVDs.[11] At 2 hours, the IOP shot up over 50 mm Hg, but at 4 hours, it had already come down to 35 mm Hg, and at 6 hours below 25 mm Hg.

The osmotic characteristic of an HA product has an effect on the state of corneal hydration. In the study by Mac Rae et al., the preparations of 1% NaHA, 0.4% methylcellulose, or 20% CS were nontoxic to corneal endothelium. However, 20% CS caused marked decrease in corneal thickness, probably because of its hypertonicity.[10]

6.1.1.4   Cohesiveness, Dispersiveness, and Adaptiveness

Intuitively, a viscous material is supposed to possess good coating property, as it is more likely to form a thick layer over the tissue, thus providing good protection. This thinking may be oversimplified when the subject is an OVD, where the concept of viscosity per se is not adequate for explaining the product attributes of coating ability under the condition of phacosurgery, and here the concept of relative cohesiveness comes into play.

Cohesiveness is best envisioned by looking at the two methods used to measure it. They are the dynamic aspiration method[12] (gravimetrically), and the cornea retention method,[13] both developed by Poyer, Chan, and Arshinoff.

Poyer et al. quantitatively determined the cohesion of ophthalmic viscoelastic agents using an *in vitro* method based on dynamic aspiration kinetics.[12] Five HA-based viscoelastics were tested: Healon GV (1.4%), Provisc (1.0%), Healon (1.0%), Amvisc Plus (1.6%), and Viscoat (HA 3.0%–CS 4.0%). Certain quantities of the viscoelastics samples were weighed in the containers. Then a calibrated vacuum was applied for 2 seconds to a pipette tip that aspirated the samples. The percentage of viscoelastic samples aspirated was plotted against vacuum pressure (100–700 mm Hg). The slopes of these curves indicate the relative cohesion (cohesion–dispersion indices) of the viscoelastic samples. The removal of the dispersive OVD showed a pattern represented by a uniform, shallow curve, whereas the cohesive OVD showed a sharper, distinct S shaped curve (a cohesive OVD displays a "break point" of the vacuum, where bolus-removal begins. The "slopes of the steepest portion of each curve" represent the degree of cohesion). The indices derived from the slopes were Healon GV (72.3), Provisc (46.0), Healon (31.2), Amvisc Plus (21.4), and Viscoat (3.4). Apparently, materials with higher cohesiveness (higher cohesion–dispersion indices) are easier to remove than materials that are more dispersive (lower cohesion–dispersion indices).

The second method was to measure the retention of viscoelastic agents on corneal endothelial cells (color staining retained). Poyer et al. quantitatively labeled confluent rabbit corneal endothelial cells with neutral red dye[13] and covered them with five ophthalmic viscoelastic agents: Healon, Provisc, Amvisc Plus, Formulation A (a dispersive, nonproteinaceous, synthetic polymer), and Viscoat. After irrigation and aspiration, with fluid turbulence similar to that encountered in phacoemulsification surgery, the cells were treated with an acidified ethanol solution to extract the dye from the cells left without a viscoelastic cover. The extracted dye was measured by spectrophotometry. The retention value (the percentage of cells with viscoelastic retained on the surface) was calculated as follows: Healon, 7; Provisc, 16; Amvisc Plus, 17; Formulation A, 55; and Viscoat, 90. The results of this experiment indicate that cohesive viscoelastics are readily removed from the cells, whereas dispersive viscoelastics are highly retained.

Arshinoff explained in his 1998 review[14] of viscoelastic materials that "cohesion allows a viscoelastic to be removed as a single mass, by aspiration or irrigation at the end of the surgical procedure." The benefit of high cohesiveness is that it allows nearly complete removal of the viscoelastic material and therefore limits the degree of elevation of IOP postoperatively. The high cohesion, however, may also cause too much of the viscoelastic to be aspirated out of the anterior chamber too rapidly, when certain amount of viscoelastic need to be retained throughout the procedure of phacoemulsification.

The opposite of cohesiveness is dispersiveness. A less cohesive viscoelastic is considered more dispersive, and vice versa. Viscoelastics are injected into the eye as a bolus. There is also clinical benefit when a viscoelastic material is well dispersed in the anterior chamber and well retained on the corneal endothelium. Such dispersiveness keeps adequate coating material to protect the cornea, when the rest of the material is removed during phacoemulsification.

When comparing the coating ability of Viscoat, Healon, and Amvisc, it should be pointed that it is not just the viscoelastic properties of those materials that make the difference in their clinical performance. Note that the total polymer concentration in Viscoat is 70 mg/ml (30 mg/ml NaHA and 40 mg/ml CS), which is several times higher than those of Healon (10 mg/ml NaHA) and Amvisc (12 mg/ml NaHA). The coating ability seems to be related to the sheer mass available to cover the tissue surface involved as well.

High MW contributes a great deal to the networking of HA, and therefore the cohesiveness of the viscoelastics, whereas low-MW materials such as Viscoat, composed of chondroitin sulfate (MW 55,000) and low-MW NaHA (0.5 million), have weaker intermolecular interaction. The low-MW polymer has to be in a concentrated form to be viscous. This is the way dispersiveness relates to low MW and a high concentration of OVDs.

What would happen if the concentration of a high-MW HA such as Healon continues to increase by 50%, 100%, or more? Dick et al. conducted a study that attained the following results. The mean viscosity at zero shear rate was 243 Pas for Healon (10 mg/ml), 2451 Pas for Healon GV (14 mg/ml), and 5525 Pas for Healon5 (23 mg/ml). (Healon5 also had the highest viscosity and elasticity when exposed to low and high shear rates). The mean relaxation times were 21, 83, and 88 seconds, respectively. Relaxation time is determined by the intersection of G'/G", and it reflects the duration of space-occupying and the effectiveness of maintaining anterior chamber depth.[3]

One of the most important characteristics of Healon5 is that this highly viscous and elastic formulation adapts to the continuous high shear rates and turbulence generated by phaco power and becomes dispersive by being fragmented. That is why such property is called viscoadaptive and also called pseudodispersive. The fragmentation of the OVD facilitates the forming of a cavity with an outer retentive shell, which is protective.

Arshinoff and Wong further examined the mechanism of such viscoadaptive behavior of OVDs by analyzing their rheological behavior.[15] More specifically, the researchers studied to what extent Poiseuille's law can actually be applied to each OVD class. Poiseuille's law describes the relations of the variables in determining the flow of a liquid. It states that the flow rate of a liquid (the rate of aspiration of HA) is determined by the radius of the aspiration port (directly proportional to the fourth power), the length of the aspiration port (inversely proportional), the pressure gradient at the aspiration port (directly proportional), and the viscosity of the OVD under the "aspiration flow conditions" (inversely proportional). Simply put, other things being equal, the aspiration rates should be determined by the viscosity of the material. In other words, the more viscous the OVD, the slower the flow rate is supposed to be. Nevertheless, this may not always be the case for HA-based OVD. According to the study by Hutz et al., the more cohesive and viscous Healon was removed much faster than the more dispersive but less viscous Viscoat.[16]

Poiseuille's equation is only applicable to noncompressible Newtonian fluids. Though being noncompressible, HA is known for its non-Newtonian behavior. The major difference in the behavior of OVDs in their aspiration, however, is not so much in the pattern of flow in the tube as in the consistency and ease with which the material is taken into the tube in the first place. Dispersive and adaptive alike, the material is removed by first breaking apart to smaller pieces, and the rate-limiting factor is often the speed with which those small fragments can be generated (often under vacuum) and enter the aspiration tip. The difference, however, is that the viscoadaptive OVDs (high MW and high concentration HA) with zero-shear viscosity as high as 7 to 18 M mPas is rigid and solid-like, therefore requires much stronger breaking force than what is needed by viscodispersive material.

Because of the nature of dispersive or cohesive OVDs, the aspirating port is not in constant contact with the material, resulting in intermittent aspiration. This is the main reason for the "deviation of the aspiration dynamics of dispersive and viscoadaptive OVDs from Poiseuille's law," as indicated by Arshinoff et al.[15] "Once the aspiration port grasps the cohesive OVD mass, it deforms and scrolls into the port; the aspiration rate can now be determined fairly accurately by Poiseuille's law." In the tube, a viscoadaptive OVD flows as one piece and is removed rapidly and completely.

6.1.1.5   Rationale and Benefit of Combination Viscoelastics

Can one recipe fit all? At least there has been no lack of effort in developing such products. Amvisc Plus, designed to strike a balance between chamber protection and easy aspiration (i.e., dispersiveness and cohesiveness), is formulated by using medium-range-MW HA (1.5 million Da) at about 16 mg/ml. Healon5 has been developed to provide a product with the advantages of both cohesive and dispersive viscoelastics. However, the trend seems to favor the dual products of a dispersive and a cohesive viscoelastic agent in the so-called dispersive–cohesive soft-shell technique,[17] which uses the two types of viscoelastic agents sequentially.

The following is the sequence of the process in cataract surgery and insertion of IOL, based on the description of Arshinoff[17]:

1. A lower-viscosity, more dispersive agent (e.g., Viscoat) is first injected to form a mound on the surface of the lens.

2. A higher-viscosity, more cohesive agent (e.g., Healon GV) is then injected into the posterior center of the first agent, which is pushed to form a smooth layer against the corneal endothelial cells.

3. During the phacoemulsification, the cohesive agent rapidly leaves the eye, and the dispersive layer in direct contact with corneal cells remains largely intact during phaco and irrigation/aspiration procedures.

4. After the removal of nucleus and cortex, the two agents are injected in reverse order, and the cohesive agent is injected first to keep the depth of the anterior chamber and stabilize the iris and capsule.

5. The dispersive agent is then injected into the center, where the incoming IOL will go through with little resistance.

6. Finally, both the cohesive agent and the dispersive agent wrapped inside are aspirated together at the end of the procedure.

In summary, the soft-shell technique maximizes the advantages and minimizes the disadvantages of both dispersive and cohesive agents.

A certain incidence of bullous keratopathy following cataract surgery may be the result of phacoemulsification now being performed in patients with hard lens nucleus. Miyata et al.[18] evaluated the soft-shell technique in cataract surgery of patients with hard lens nuclei. They compared the soft-shell technique using a combination of Viscoat and Healon with the control group using Healon alone. It was found that the central corneal thickness was significantly greater in the control group than in the group that had the combination of Healon and Viscoat in the soft-shell technique.

Behndig and Lundberg[19] also evaluated protective effect of the soft-shell technique in a three-arm clinical study of phacoemulsification. The first group received Healon GV only. The second group was given Viscoat/Healon GV combination, and the third group a Viscoat/Provisc combination, both using the soft-shell technique. The authors examined the transient corneal edema by assessing corneal thickness and found that at 5 hours, 24 hours, and 1 week, the mean increase in corneal thickness was significantly greater in the first group than in the two groups using the variation of the soft-shell technique. The above two examples are indicative of the extra benefit that the soft-shell technique may bring to protect the corneal endothelial cells.

Arshinoff took the concept of soft shell further to include the use of the highest-viscosity product Healon5 in combination with a lowest-viscosity, nonpolymer-based solution such as BSS.[20] This combination has led to the ultimate soft-shell technique. The technique is particularly useful during capsulorhexis (making a continuous circular tear in the anterior capsule), because the resistance to needles and forceps for capsule tearing can make the operation difficult if only viscoadaptive material is used. Conceivably, the resistance would be minimal when BSS is used at the area immediately in contact with the capsule. This is feasible for the cataract surgery, when

BSS is used in conjunction with Healon5, which can block the incision so that BSS does not leak out.

6.1.1.6   *Removal of Viscoelastics Postsurgery*

To avoid postoperative intraocular pressure surge, it is important to remove, as much as possible, the OVDs used during cataract surgery. As discussed before, the concepts of cohesiveness, dispersiveness, and adaptiveness have been developed on the technical basis of OVD removal, and these properties are relative. Strictly speaking, the so-called dispersive OVDs are fairly cohesive as compared to the real dispersive solutions such as water and saline. All polymers are somewhat cohesive because of chain interactions between molecules.

Viscoadaptive OVDs provide a new challenge to their removal postsurgery. It requires turbulence (generated by high-vacuum stress) to fracture the rigid material and to allow the small fragments to scroll around obstacles and into the irrigation/aspiration (I/A) port. The OVD fragments may be trapped behind the IOL (and on the sharp edges of modern IOLs), where turbulence is limited unless the I/A tip is placed behind the IOL. Therefore, like dispersive OVDs, the aspiration of viscoadaptives is relatively slow, because of their "frequent escape from the clutch of the aspirating port during attempted aspiration."[15]

Two major methods for removal of viscoadaptive OVD have been developed and compared.[21–23] They are rock 'n'roll, in which the OVD is removed only from the anterior chamber, and behind-the-lens — also called the two-compartment technique — in which the OVD is removed from behind the IOL in the capsule as well as from the anterior chamber. According to Zitterstrom et al.[21] in the rock 'n'roll method for OVD removal, the I/A handpiece, applying gentle pressure on the IOL optic, "was rotated first to one side with the flow directed into the bag and then to the other side using the same maneuver" whereas in the behind-the-lens method, the OVD was removed by going behind the IOL, where the "flow was not engaged until the I/A tip was positioned behind the IOL optic with the aspiration port facing the cornea". The two-compartment technique removes the OVD more quickly. In the study conducted by Tetz and Holzer, it took on average 18 seconds to remove Healon5 using two-compartment technique, and 44 seconds using the rock 'n roll technique.[23] This is consistent with the *in vitro* study result that "with some OVDs, I/A from behind the IOL is necessary for complete removal."[24] The study by Zitterstrom et al. also showed that the 5-hour postoperative IOP increase was +9.9 mm Hg in the rock 'n'roll group ($n = 79$) but only +6.5 mm Hg in the behind-the-lens group ($n = 80$), an indication that the latter is more efficient.[21]

The benefit and utility of an OVD is invariably related to the condition and surgical technique with which it is used. As Mamalis commented: "ophthalmic viscosurgical devices are no longer a simple material that maintains space or coats ocular tissues," they have become an integral part of new phacoemulsification technologies.[25] This has led to a paradigm shift in our thinking about viscoelastic materials, which serve the need of a surgery and are being perfected by it as well.

6.1.1.7   Pharmacoeconomic Considerations

Before addressing the pharmacoeconomic aspect of the HA-based OVDs, it is interesting to notice that the way the effectiveness of OVD was established is quite unique, as we revisit the early history of FDA approval of OVDs containing HA, such as Healon. Clinical outcome was not considered in the PMA summary documents as the critical criteria because clinical outcome can be determined by many other factors. In the summary of PMA 810031, for Healon, the FDA commented that "the success rate of the surgery in which Healon was used is not an appropriate criterion for the effectiveness of Healon because the patient's final outcome is affected by preoperative severity, surgical techniques, general health of the patient, and other therapeutic measures." As a surgical aid, Healon is intended to help the physicians perform the surgery through its desirable characteristics in viscosity and transparency, "which allow the device to remain in the surgical area or to hold tissue in place during and after surgery." In this case, the FDA considered the "investigator's experience or evaluations of Healon's usefulness" as "valid scientific evidence of Healon's effectiveness." As compared to its alternative, low-viscosity BSS, the ability of OVD to aid surgery as a result of its viscosity was so obvious that the effectiveness could, in large part, be established through the direct observation and experience of the clinicians (the protection of endothelial cells by OVDs was treated as a matter of safety, for which HA-based OVDs demonstrated an excellent record, as described in Chapter 5). This explains why HA-based OVD quickly gained popularity among the surgeons, once they had the opportunity to use it.

Is OVD effective from a pharmacoeconomic standpoint? Kiss et al.,[26] in Austria, compared the protective effect on corneal endothelial cells of a lower-cost hydroxypropyl methylcellulose 2% (Ocucoat) and higher-cost Viscoat and concluded that there was no difference between the groups in acute postoperative corneal edema and endothelial cell morphology 3 months after phacoemulsification, and eyes receiving the expensive Viscoat had only marginally faster recovery of corneal swelling.

Although there are some valid points in that argument, the convenience that a transparent and viscous HA and CS containing OVD bring to the surgeons should not be underestimated. If an OVD containing HA can contribute to even only a "marginally faster recovery," it may already be worth the extra cost. Cataract surgery, after all, is a critical and nonrecurring surgery, and its success is of utmost importance to any patients under any conditions. Some ophthalmic HA products are expensive, but the price of HA for ophthalmic use has been in decline over the years because of competition. In addition, the cost of HA is somewhat a myth. Only a small amount of HA is needed for a cataract surgery, usually less than 1 ml, and the cost of that amount of HA is probably no more than the cost of the syringe in which HA is filled. As more competitors enter the market, the price of HA viscoelastic for ophthalmic use should not be the major hindrance to its wide application in the future.

6.1.2   Vitreous Substitute

The vitreous humor of the eye (i.e., the vitreous body) is a transparent gel occupying an intraocular space from the retina at the rear of the lens to the ciliary body at the front. The retina comprises two layers, the receptor layer containing photosensitive cells and the layer adjacent to the choroid consisting of pigment epithelial cells.

When a hole has been torn in the retina and the fluid penetrates through the receptor layer, this may result in a separation of the two layers. The rupture of retina can be closed by a coagulation (such as cryo method) of the detached retina layers. For the coagulation to work, additional mechanical force is often needed. Two of the common approaches for providing this force are impressing an inward buckle from the outside on the sclera and choroids and injecting materials (gases, liquid, and gel) to increase the volume of the vitreous humor and thereby increase the pressure exerted by the vitreous humor on the retina.

Silicone oil continues to be used in retinal detachment surgery, perhaps because of its physical characteristic of relatively high interfacial tension in water, which is important for such surgery. Silicon oil needs to be removed within 4–6 weeks because of the long-term complications of cataract, keratopathy, and glaucoma. Perfluorocarbon liquids have found its use in short surgical procedures for the manipulation of the retina, but it is unsuited for longer-term use. Leaving it in the eye beyond 1 week may cause retinal disorganization.[27] A combination product with HA and collagen was attempted by Balazs and Sweeney in the 1960s.[28] After all, these two biopolymers are the natural components of vitreous body, and a combination product (called reconstituted vitreous in the paper) would make much sense. The major problem of the reconstituted vitreous, however, turned out to be its lack of transparency, in sharp contrast to the formulation with HA alone.

In vitreous surgery, HA also demonstrated excellent tissue protection. Nevertheless, the goal of using HA as a long-term tamponade against the detached retina has not been met. The main reason for the disappointment is that the residence time of HA in the vitreous chamber is too short to serve as an effective tamponade. Crosslinking of HA was thought to be promising, at least in concept, to solve this problem. Nevertheless, after much effort for many years, no crosslinked HA has been able to prove its safety and effectiveness as a clinically useful tamponade to treat retinal detachment. As mentioned in Chapter 5, a chemically crosslinked HA may not necessarily be as biocompatible as native HA. This is especially true for tissues as sensitive and delicate as those of the eye. Nor would the other ideal characteristics of native HA formulation such as transparency and coherency be likely maintained, when a high degree of crosslinking of HA is attempted to increase its *in vivo* residence time.

In the 1970s, Pruett et al. experimented with the use of native HA formulation as vitreous substitute, in conjunction with scleral buckling, to repair rhegmatogenous retinal detachments: "Scleral buckling entailed scleral undermining, treatment of the of the choroid and retinal pigment epithelium beneath retinal breaks by diathermy or cryoapplications, solid silicon rubber implants with overlying nonabsorbable mattress sutures placed in scleral flaps, and encirclement with a silicone rubber band."[29] HA was injected into the vitreous space following the release of subretinal fluid and the development of buckling.[29] The HA (Hyvisc) was extracted from rooster comb and prepared as an isotonic solution containing 0.9% w/v HA. Between 3 and 4 ml of HA was injected, after scleral buckling, into the vitreous space, with the needle tip positioned as close to the optic disc as possible. It was hoped that the viscous vitreous substitute would mechanically tamponade the retinal breaks against the treated and buckled areas. For the 73 patients undergoing the procedure of internal HA tamponade and scleral buckling, the short-term results at 1 week were favorable, with 47% reattachment rate (of retina). However, the vitreoretinal traction progressed with time in many eyes. At 1 month, the total reattachment rate was down to 21%, and at 6 months the reattachment was maintained in only 16% of the patients.

Buratto et al.[30] conducted research to test the residence time of OVDs in rabbit eyes. In the filled vitreous chamber, HA was retained for a week, and by the end of second week, only 62% HA was left. This was slightly better than the hydroxypropyl methylcellulose (HPMC) formulation. Only 60% of HPMC remained after 1 week, and 38% after 2 weeks.

The residence time of HA formulation is too short to accomplish any meaningful internal tamponading. Buratto et al.[30] concluded that "the use of viscoelastics as tamponading devices does not seem to be practical" but recognized that the main utility of HA-based OVD in vitreous surgery is to "allow manipulation of posterior segment tissues with a precision and care that could not be achieved otherwise while ensuring and improving the visibility of procedures." The benefit for HA as a surgical aid in certain vitreous surgery was also recognized by Pruett et al.[29] two decades earlier, which included the improvement of the visualization during surgery, better preservation of corneal epithelium, and so on.

Attempts were also made to increase the residence time of HA by chemical crosslinking. Examples can be found in patent publications of the U.S. Patent and Trademark Office. In USP 4,716,154, Malson et al. described the "gel of crosslinked hyaluronic acid for use as a vitreous humor substitute." The crosslinking of HA was by bifunctional (e.g., butanediol diglycidyl ether BDDE) or polyfunctional epoxide, or by a corresponding halohydrin, epihalohydrin, or halide. The chemistry has been described in Chapter 2. The patent summed up the ideal characteristics of material to function and perform as a vitreous substitute, that is, to possess a high degree of transparency and about the same refractive index as the vitreous humor; to be a substance that can be applied easily, for example, by injection through a fine needle tip; to be capable of controllable swelling; and to act as a support during a prolonged period of time therefore not easily degraded or decomposed. The patent claimed that the gel prepared by the patented technology "has properties in very close agreement with those set forth above as characteristics of an ideal vitreous humor substitute." The fact that no crosslinked HA has gained approval in the U.S. for vitreous replacement more than 20 years after the approval of native HA for other ophthalmic indications suggests the complexity of this project. A "close agreement" is probably not adequate to make the product safe and effective. The challenge to fully developing a truly useful HA-based vitreous substitute is yet to be met.

6.2 Injections for Osteoarthritic Knees

Since 1997, five HA-based products have gained FDA approval for the treatment of pain of osteoarthritis (OA) of the knee. They are Hyalgan, Synvisc, Supartz, and two recent additions of Orthovisc and Nuflexxa. The marketing history of these HA products for knee injection in the U.S. has been relatively short, but the controversy in the theory and practice of this particular therapy runs deep. The use of HA in ophthalmic surgery has gained universal acceptance in the medical community, whereas the clinical value and utility, as well as the mechanism of action, of HA-based formulations for OA of the knee remain contentious issues for debate. How effective is local injection of HA in alleviating the pain of the arthritic knee? To the extent it works, what is the mechanism behind it? What product characteristics are truly contributing to the therapeutic effect the products may have on the treatment? How important is viscoelasticity of HA for knee injection?

6.2.1 The Maze of Clinical Data in PMAs

To investigate the performance and efficacy of HA-based products in treating OA of the knee, one had to first determine the standard or reference point to which the comparison could be made. A review of the history of market approval of Synvisc and Hyalgan in the U.S., the first two products of this kind, indicated that the primary criteria for comparison, agreed on by the U.S. regulatory agency (FDA) and the sponsors, was the buffer solution without HA solute. Note that the clinical study design of those products was based on the assumption that the injection of buffer solution, or arthrocentesis (a procedure whereby a sterile needle and syringe are used to drain fluid from the joint), would have no therapeutic effect, and therefore could be used as negative controls.

Several clinical outcomes of the use of HA-based products in relieving pain of the OA knee were based on a variety of pain and function scores, as examined by the FDA during the regulatory reviews for approval of Synvisc (1994–1997), Hyalgan (1995–1997), Orthovisc (1996-2003), and so on. The *Summary of Safety and Effectiveness of Synvisc, P940015,* mentioned several pain scores (e.g., weight-bearing pain, night pain, painful knee movement, walking pain, etc.), and the *Summary of Safety and Effectiveness of Hyalgan, P950027,* mentioned the visual analog score (VAS) for pain after walking 50 feet. In the study of Orthovisc,[38] the WOMAC (Western Ontario and McMaster Universities Osteoarthritis Index) pain score used had a total of 25 points (five different pains in five degrees of severity based on the Likert scale). The primary measure for effectiveness was the relief of various pains in comparison with the saline injection or arthrocentesis. The differences between the HA and saline groups in favor of HA were expected to be clinically significant, and human studies were designed to have enough statistical power to detect such differences.

6.2.1.1 Strong Placebo Effect of Saline

It turned out that the expected "negative control" of saline was not negative at all, and the placebo effect of the saline injection in alleviating the knee pain was fairly significant. In both Synvisc and Hyalgan studies, saline injections demonstrated marked improvement in pain score from baseline. In most studies of Synvisc, the improvement was over 20% from baseline, and the effect lasted up to 12 weeks (*Summary of Safety and Effectiveness of Synvisc, P940015*: the 12-week saline injections pain scores were listed in its Tables 5, 7, 10, and 12). In the Hyalgan study, over 40% of the "placebo" patients had greater than 20 mm improvement over baseline (definition of success in that study) for continuous variables in the VAS, "observed at week 5 and maintained until week 26" (*Summary of P950027*, Table 3: VAS for pain at a 50-foot walk [20 mm improvement]). As a consequence, cases of significant difference between HA group and control were more exceptions than the rule. The overwhelming majority of the studies included in the 1996 FDA panel reviews (FDA, Orthopedic and Rehabilitation Devices Panel, Open Session, November 20–21, 1996) failed to show a between-group statistical significance even in one pain score. None of the studies demonstrated significant difference in all prospective outcomes.

6.2.1.2 Synvisc

Each course of treatment by Synvisc takes 3 weekly injections. Among the seven studies of Synvisc reviewed by the FDA, six were double-blind controlled studies, and only one large study, conducted in Germany (Study 3, 110 patients) by Wobig et al., showed a between-group significant difference in various pain scores.[31] One cannot help but notice that the difference in favor of Synvisc in that study was remarkable. For example, when measured by VAS, the improvement of patient-evaluated weight-bearing pain from baseline at 12 weeks was greater than 30 mm (46.5–16.4; $P < .0001$). However, the only U.S. study of Synvisc under review (Study 5), in which 94 patients were enrolled, failed to demonstrate any statistically significant difference for any pain scores at the 12-week visit. The improvement of patient-evaluated walking pain from baseline at 12 weeks, for example, was less than 2 mm (21.0–19.5). The discrepancy of such magnitude was puzzling. It was noted, however, that in the Synvisc study in Germany, patients lacked intraarticular (IA) effusions, whereas in the U.S. group, the patients had knee effusion and the control group of patients received arthrocentesis, but no saline injections ensued. Nevertheless, there is no strong evidence that this actually accounted for the dramatic difference of the results of two studies of the same product. Brandt et al. commented[32] that in the Wobig study, the treatment and control groups did not match at baseline with respect to radiographic severity. The proportion of patients in Larson grade III–IV was 28% for Synvisc, and 57% for saline control. In addition, other variables in the study end points may be relevant to the difference of the outcome. For instance, there were many different scoring systems and subcategories just for pain treatment alone, such as weight-bearing pain in the German study, walking pain in the U.S. study, and so on.

6.2.1.3   Hyalgan

Each course of treatment by Hyalgan takes five weekly injections. Hyalgan was the first HA product approved for marketing for OA indication in the U.S., and its approval was mostly based on the results of a pivotal study (495 patients) in the U.S. that compared Hyalgan with saline injection and oral naproxen. The study was also published by Altman and Moskowitz.[33] Three pieces of information were considered as key evidence from that study, supporting the efficacy of Hyalgan: first, patients receiving HA improved more with respect to pain on the 50-foot walk than placebo at week 26 (difference of VAS 8.8 mm, $P < .005$); second, more HA-treated patients (56%) than saline-treated patients (41%) had 20 mm or more reduction in the VAS from week 5 through week 26 ($P = .031$); and third, at 26 weeks, more HA-treated patients (47.6%) had slight pain or were pain-free than saline-treated (33.1%; $P = .039$) or naproxen-treated (38.9%; $P = .022$) patients.

   In the *Summary of Safety and Effectiveness of Hyalgan (P950027)*, the results of VAS for pain after the 50-foot walk test were summarized as follows: "Among study completers (HA 105, placebo 115), the Hyalgan-treated group exhibited statistically significant greater improvement in this parameter as compared to the placebo group at weeks 4, 5, 12, 21, and 26 (p < 0.05)." The magnitude of the difference at week 26 was 8.85 mm, which, although better than any other measures in this pivotal study, is still a small number in its absolute value. The reason that such difference could even reach statistical significance had to do with the larger study population (over 100 patients per arm) and the small standard deviation (8.85 mm accounted for approximately one-third of the standard deviation of the between-group difference). However, in the other two so-called "primary effectiveness parameters" (i.e., categorical assessment of pain by masked evaluators and patients themselves), the differences were not statistically significant, although slightly in favor of Hyalgan.

Felson[34] and Lo[35] challenged the validity of relying on the data of compl-eters (per protocol population). They emphasized that "an intent-to-treat analysis provides the most valid assessment of the efficacy of therapy because it preserves the randomization of the trial and does not allow the nonrandom nature of dropouts to bias trial results."[34] It was pointed out in the Altman study, the dropout rates were 26.1%, whereas the intent-to-treat analysis was described as nonsignificant (no details were provided). This raised concerns about the accuracy of the effects reported.[35]

Felson et al. went on to analyze the data of intent-to-treat analysis of four large studies[33,36–38] each with more than 200 subjects, and concluded that "HA has little or no effect in the treatment of OA compared with placebo." Brandt also commented with regard to the HA for knee injection in general that "it is not clear whether improvement after HA injection is real, due to a placebo effect, or due to a regression-to-the-mean phenomenon (where patient with symptoms more severe than average are selected for IA injec-tion, and their pain would improve even without treatment)."[39]

6.2.1.4   *Orthovisc*

Orthovisc was approved by the FDA in 2004, based on three clinical trials (OAK9501, OAK9801, and OAK2001) conducted by Anika Therapeutics between 1996 and 2003 (Ortho-Biotech Orthovisc High Molecular Weight Hyaluronan, *Full Prescribing Information*, February 2004). It is quite interest-ing to look at the journey Orthovisc made to gain the regulatory approval from the FDA. The first pivotal clinical study of Orthovisc (OAK9501) was conducted from 1996 to 1997, at 10 sites in the U.S. The study of Orthovisc, like that of Synvisc and Hyalgan, showed a very strong placebo effect of saline injection. Take one of the main clinical end points WOMAC pain score as an example — the improvement of pain from baseline in the saline group accounted for nearly 80% of the improvement of the Orthovisc group toward week 27.[38]

In a post hoc analysis of a subgroup that only included those patients in the Orthovisc study with baseline contralateral knee pain at no greater than 12 (Orthovisc group 66, saline 69), the results showed a statistically signifi-cant between-group difference. At 27 weeks, the improvement from baseline favored Orthovisc by approximately 1 WOMAC point. However, the intent-to-treat population (Orthovisc 114, Saline 112) failed to demonstrate such statistical significance. Although there appeared to be a trend in favor of the Orthovisc numerically over the study period, the difference at 27 weeks was only 0.3.[38] The FDA requested additional data for review.

Further analysis showed that the pain score for the contralateral knee was inversely related to the magnitude of improvement of the pain score of the HA-treated knee. In other words "moderate to severe contralateral knee pain may mask the robustness of the pain response in the index knee" after the HA treatment.[38] Anika concluded that the new study had to solve the "confounding" issue ("Anika Therapeutics Says FDA Requires Additional Clinical Data for Approval of Orthovisc," *Business Wire*, October 19, 1998: "the [FDA] letter stated that the data failed to demonstrate effectiveness according to the study's pre-determined success criteria. . . . The Company believes that pain in the untreated knee may have had the effect of confounding the study results").

The second Orthovisc study (OAK9801) included 385 patients and was completed by the end of February 2000. The control group was the same as the first study — saline injection. To date, public information with regard to this second study has been scarce. A brochure entitled *Orthovisc Full Prescribing Information* by Ortho Biotech and Anika Therapeutics mentioned that the study "involved bilateral treatment" that "confounded the assessment of effectiveness of the OAK9801." It would appear that the very reason to change the study to bilateral design (injecting HA also into the painful contralateral knee) in the second study was supposed to avoid the confounding effect of the pain of the other knee. It was not clear how the new study design, intended to avoid confounding by using bilateral injections, resulted in causing more confounding. The company announced in 2000 that the study "did not show sufficient efficacy" ("Anika Therapeutics Reports Unfavorable ORTHOVISC Clinical Trial Results," *Business Wire*, May 30, 2000: "an initial analysis of results from its recently completed Phase III clinical trial of ORTHOVISC. . . did not show sufficient efficacy to support the filing of a pre-market approval (PMA) application. . . . the study did not achieve the primary efficacy end-point, a measure of pain reduction").

Some major changes were made in the design of the third study of Orthovisc (OAK2001), presumably to increase the probability of success in showing the efficacy. First, this third study replaced saline injection with arthrocentesis in the control group, hoping the placebo effect of the latter (without saline) would not be as large as the former (with saline). Second, four injections of HA were performed instead of the three in the previous two studies. Third, instead of direct comparison of the improvement from baseline of the treatment and control groups, as done in previous studies, the primary effectiveness criteria became "proportion of patients achieving a 20% improvement from baseline in the WOMAC pain score in conjunction with a minimum absolute improvement of 50 mm from baseline in the WOMAC Pain Score, and a 40% and 50% improvement from baseline in WOMAC Pain Score. . . for the index knee" (at four assessment points up to 22 weeks). In addition, the WOMAC pain score in the study was on a continuous VAS, not on the five-grade Likert scale. Even with such changes in design and outcome measures, the primary analyses for study I and study III did not show statistical significance when analyzed individually. It was not until a subgroup analysis of patients "who had Kellgren-Lawrence radiographic grades of II or III at baseline and WOMAC pain in the contralateral knee of <175mm (out of 500)" did the significance begin to emerge. Furthermore, even the newly defined end point as described in the third point above was only partially met; that is, significance was only found in subgroups with 40% and 50% improvement, but not in the subgroup with 20% improvement and improvement of 50 mm from baseline.

There are two kinds of WOMAC pain scores. Orthovisc study (I) used WOMAC Likert Score, with 25 total points. Study (III) used WOMAC VAS with a maximum of 500 mm. A contralateral knee pain of 12 in Likert scale is equivalent to 240 mm pain in VAS of a 500-mm scale. There is no clear logic why excluding patients with a pain score of contralateral knee of 240 mm or more was not acceptable in study I (OAK9501) but excluding patients with pain score of 175 or more in the contralateral knee was allowed in study III (OAK2001), as both studies were post hoc analysis. Other things being equal, the subgroup in study III for effectiveness analysis actually deviated more from its whole intent-to-treat group by adopting a more restrictive exclusion criterion (the contralateral knee pain at baseline ≥175 mm) as compared to the case of Study I (OAK9501).

Orthovisc studies in a sense substantiated the general findings about the marginal efficacy of the HA-based products for treating the pain of the knee of OA. Had it been more effective, the demonstration of efficacy would have been more straightforward. In the meantime, regulatory approval is not an exact science. It depends, to a great extent, on the judgment by the agency under a particular situation. Sometimes, the FDA advisory panel can have a critical influence on the decision of the FDA as well. In the case of the Orthovisc review, however, there seemed to be another decision influencer at work; that is, the quantity of the data. Often, the persuasion power of only one or two limited clinical studies is just not adequate. As more data became available through new studies, so did the general picture of the similarity in the safety and efficacy profile of Orthovisc become clearer, versus products with same indications approved earlier. This could have a positive effect on overcoming the hurdle from the literal interpretation of the prospective success criteria — a major factor that had seemingly prevented U.S. approval of Orthovisc in the past.

6.2.1.5   *Supartz*

The clinical trials of Supartz reviewed by the FDA for approval included five studies conducted in Australia, France, Germany, Sweden, and the United Kingdom (*Summary of Safety and Effectiveness Data, Supartz, P980044*). They were randomized, blinded, placebo controlled clinical trials. A total of 619 patients received Supartz injections, and 537 patients in the control groups received phosphate buffered saline (PBS) injection. The treatment regimen consisted of five weekly injections in all studies. Patients in these studies (including those injected with the control) received arthrocentesis of the knee before an injection of Supartz or phosphate buffered saline. The French study included an additional treatment arm: three Supartz injections followed by two injections of the control per patient.

The clinical performance of Supartz in relieving pain and improving function of the knee was evaluated. In the Australian study (108 Supartz/115 Control), WOMAC index of pain, stiffness, and disability of the Supartz and the control were compared, and the results in all three categories favored the treatment group — Supartz — but only the pain and stiffness showed significance. Another algofunctional index Lequesne score (a combination score of pain, walk, and activities for knee OA)[40,41] was used to compare the results in all five clinical studies. Only the Australian and U.K. studies showed between-group significance.

The pattern of clinical performance of Supartz is very much like that for the other HA products for knee injections. There is a general trend of Supartz performing better than the saline solutions, but in terms of predetermined clinical end points, there were hits and misses in reaching between-group statistical significance.

Although Supartz entered the U.S. market after Hyalgan and Synvisc, it was the first HA-based product commercially available for human use in treating OA. According to Smith & Nephew, by the end of 2004, "more than 102 million Supartz injections have been sold worldwide since its introduction in 1987" (http://www.smith-nephew.com).

6.2.2 The Big Picture — Metaanalysis

How effective is the therapy of HA injection for knee OA? For every paper that endorses such treatment, there appears to be another refuting it. Instead of limiting one's view only on selected, individual cases, it is important to examine the universe of clinical evidence in its entirety, and a metaanalysis would serve such a purpose. Lo et al. conducted such analysis and published the results in 2003's JAMA.[35] In the study, extensive search of HA (hyaluronic acid, Hyalgan, Synvisc, Artzal, Suplasyn, BioHy, or Orthovisc) and OA (osteoarthrosis or degenerative arthritis) in Medline (1966–February 2003), Cochrane Controlled Trials Register, special medical journals (*Arthritis and Rheumatism, Osteoarthritis and Cartilage*, and *Journal of Rheumatology*), and abstracts of scientific meetings (American College of Rheumatology Annual Scientific Meeting and the Osteoarthritis Research Society International Meeting, 1986–2002) was performed. Effort was made to include unpublished data as well.

The search criteria for human clinical studies of IA injection of HA in the treatment of OA include randomized, single-, or double-blind controlled trials against a "placebo"; weekly administration for at least 3 weeks; pain assessed and reported using one of the outcome measures recommended by the Osteoarthritis Research Society (Pain Outcome Hierarchy: 1. Global knee pain score [visual analog or Likert scale], 2. Knee pain on walking [visual analog or Likert scale], 3. WOMAC osteoarthritis index, 4. Lequesne index, and 5. Knee pain during activities other than walking [visual analog or Likert scale])[42]; a minimum follow-up time of 2 months; and a dropout rate of less than 50%.

The effectiveness data pertaining to pain relief were extracted from measures of change from baseline at 2–3 months if available; otherwise, they were extracted at 1–4 months after the first IA injection. An analysis was considered intent-to-treat if there were no dropouts. Some intent-to-treat analyses did not provide extractable data but were characterized as such by their investigators. Seven analyses included the extractable data and were provided to Lo et al. at their request for metaanalysis. Those were the studies by Dahlberg,[43] Creamer,[44] Sala,[45] Carrabba,[46] Wobig,[31] Pham,[47] and Jubb.[48]

The key effectiveness parameter used by Lo et al. in their metaanalysis is the effect size, defined as the pain improvement divided by the pooled standard deviation (SD).

$$(Pain_{HA} - Pain_{placebo})/SD$$

The degree of clinical significance as represented by the effect size can be approximated as follows: 0.2–0.5 is considered the range of a small effect size, equivalent to the effect of nonsteroidal antiinflammatory drugs over that of acetaminophen in OA trials.[49] A large effect size is at or above 1.0, reminiscent of a total knee replacement that has an effect size between 1.0 and 1.8.[50] The pooled effect size for the treatment of knee OA with HA injection versus buffered saline, according to the analysis by Lo et al., was 0.32 (95% confidence interval, 0.17–0.47; $P < .001$), indicative of a small effect.[35]

On the basis of the findings of their metaanalysis, Lo et al. concluded that

1. Intraarticularly administered HA, at best, has a small effect above what a buffer solution can provide in the treatment of pain of knee OA.

2. The injection of saline control contributes approximately 80% of the treatment effect of IA HA ["The pooled change in the hyaluronic acid arms was 0.82 SD units and the pooled change in the placebo groups arms was 0.65 SD units, suggesting that the intra-articular placebo effect accounted for 79% (0.65 SD units/0.82 SD units) of the efficacy of intra-articular hyaluronic acid"[35]].

3. The 95% confidence intervals of the effect size of most trials had included an effect size of zero, consistent with no effect or a small effect of this treatment.

4. The effect of HA treatment is equivalent to the effect of nonsteroidal antiinflammatory drugs over that of acetaminophen.

5. There was publication bias without which the actual effect could have been even smaller; 17 of 22 trials were industry sponsored. The lack of any trials with a negative effect size is also an indication of publication bias.

6. The two clinical studies in Germany by Wobig and Scale[31,51] were considered as "outliers" within the context of the meta analysis.

Clinically speaking, the effect size of these two trials (1.55 and 1.76) would be comparable to a total knee replacement, which does not seem realistic.

7. An alternative explanation to strong placebo effect could be the possible special effect of HA in a particular subgroup. However, none has been consistently identified as such a subgroup across multiple trials.

Since the publication of the metaanalysis, a new round of correspondence has been exchanged between Lo et al.[52] and Moskowitz et al.,[53] in which both sides further elaborated their opinions. There were also occasions of misunderstandings. For example, when Lo et al. stated that the effect size of HA versus placebo was similar to that of nonsteroidal antiinflammatory drugs versus acetaminophen, they were expressing doubt about the efficacy of HA. However, such a statement was taken as an approval or affirmation of the efficacy of the HA-based products. The debate continues.

6.2.3   Recommendations by European League against Rheumatism

So far the effectiveness of HA-based IA products for knee OA has been evaluated based mainly on the comparison between HA and a designated control group — mostly saline injections. To put the role and value of HA therapy in perspective, it is best to view the HA treatment of knee OA within the context of the whole panoply of viable treatment choices. Such information and opinions are often available through the organizations of health professionals in the specific field. Particularly important and relevant to the HA use in OA treatment is the recommendation by the European League Against Rheumatism (EULAR).

EULAR Recommendation 2003 took an evidence-based approach to the management of knee OA. It provided insight on the subject, as well as copious information.[54] The recommendation was made by the members of the expert committee on knee OA (19 rheumatologists, 4 orthopedic surgeons) and one research fellow each from 13 European countries.

It should be noted that knee OA is associated with symptoms of pain and functional disability, and the EULAR recommendation is based on the effect of interventions on alleviating both symptoms. So far, the limited evidence of therapeutic effect of IA injections of HA has been mainly from the pain relief. There was no clear evidence of the effectiveness of HA on joint function of OA patients in the clinical studies submitted to the FDA for marketing approval (including Hyalgan, Synvisc, and Orthovisc), except that only one of the five trials of Supartz, conducted in Australia, showed significance for better mean reduction of knee stiffness from baseline than placebo over week 6 to week 18 (*Summary of Safety and Effectiveness Data, Supartz, P980044*).

Second, the recommendation defined three categories of treatments for OA: nonpharmacological treatments (e.g., education, exercise, and lifestyle changes), pharmacological treatments (e.g., paracetamol, nonsteroidal anti-inflammatory drugs, topical treatments), and invasive interventions (e.g., IA injections, lavage, arthroplasty). The IA injections of HA belong to the invasive intervention, in the same group as arthroplasty, although in fact IA injection is not nearly as invasive as knee replacement.

EULAR weighs heavily on the totality of the evidence (metaanalysis), the effect size of the difference between treatment and placebo, and the randomization of the study. Their recommendations also take into account expert opinions on other aspects of the intervention, such as side effects, practicality of delivery, and economic considerations.

The following are the definitions of levels of evidence, in descending order of their recognized strength, of a clinical study:

Level 1A: Metaanalysis of randomized controlled trials (RCTs)

Level 1B: At least one RCT

Level 2A: At least one controlled study without randomization

Level 2B: At least one quasi-experimental study

Level 3: Descriptive studies (comparative, correlation, or case–control)

Level 4: Expert committee reports or opinions or clinical experience of respected authorities.

Effect size, as mentioned before, "reflects the magnitude of difference between two groups in standardized terms and is free of units." Clinically, an effect size of 0.2 is considered small, 0.5 is moderate (and would be recognized clinically), and greater than 0.8 is large.[54] Effect size is an important part of the consideration by EULAR in its recommendation of a therapy.

The strength of EULAR recommendation was graded from A to D, with A as the highest ranking. As seen from Table 6.1, IA HA ranks right with herbal remedies, acupuncture, or lavage, and behind glucosamine and nonsteroidal antiinflammatory drugs. The invasive nature of IA injection, the

TABLE 6.1

Examples of EULAR Recommendations as Related to HA Therapy

Intervention	Level of Evidence	Effect Size Range	Strength of Recommendation
Acetaminophen/paracetamol	1B	NA	A
Conventional NSAID	1A	0.47–0.96	A
Glucosamine	1A	0.43–1.02	A
Herbal remedies	1B	0.23–1.32	B
Acupuncture	1B	0.25–1.74	B
IA hyaluronic acid	1B	0.0–0.9	B
IA Corticosteroid	1B	1.27	A
Lavage/tidal irrigation	1B	0.84	B

NA = not available

cost of the HA treatment, and the potential side effects of such intervention (especially in light of the reports on adverse events involving Synvisc) are all factors in the equations of EULAR recommendations. Within this context, it is not surprising that the effectiveness of HA injection will have to be really striking to win a more favorable rating. So far, in an overwhelming majority of the cases, IA HA has failed to demonstrate an effect of such magnitude. It is noted that the *Indication for Use* of Hyalgan, Synvisc, and so on, is "for the treatment of pain in OA of the knee in patients who have failed to respond adequately to conservative nonpharmacologic therapy, and to simple analgesics, *e.g.* acetaminophen." It is clear that the noninvasive, more conservative therapies should be considered first. What is not clear is exactly what kind of subpopulation may respond to HA injection consistently.

6.2.4   Mechanism of Action

Similar to the controversy over the therapeutic value of HA-based products in the treatment of OA of the knee, little consensus has been reached on its mechanism of action. There are at least two different schools of thought on why HA can relieve the pain of OA knees to the extent it has demonstrated. One theory, as represented by viscosupplementation, emphasizes local and physical effect, and the other believes the main effect is systemic.

6.2.4.1   *Local versus Systemic*

The concept of viscosupplementation was introduced by Balazs et al.[55]: "The overall mode of action of viscosupplementation is the restoration of the physiological homeostasis of the joint." Homeostasis is a state of physiological equilibrium in the joint. There appear to be several biological and physiological events interrelated to the viscosupplementation:

1. Pathological synovial fluid has low MW and viscosity, the injections of elastoviscous HA gel restore the rheological property of the synovial fluid.

2. The exogenous HA may stimulate hyaluronan synthesis and restore the normal hyaluronan content.

3. Viscous HA influences the nociceptors, located in synovial and subsynovial tissue, thus relieving the pain of the joint.

4. Joints free of pain are more mobile, which leads to improved transsynovial flow and facilitates cell metabolism and removal of the debris.

5. Within such a healthy environment, the cells (hyalocytes) may now start to produce normal HA.

However, the theory of viscosupplementation is facing serious challenges. The lack of strong clinical evidence to support the efficacy of local injection over placebo, as discussed in the previous section, does not support the viscosupplementation. Another key argument against the theory points to the mere fact that HA administered intravenously (IV) in animals actually works just as well as the HA injected intraarticularly, as will be discussed in the following section.

The marketing history of the HA-based veterinary products indicated for horse joints with noninfectious synovitis in fact predated that for the treatment of OA for humans (Table 6.2). The route of administration of the five products for race horses approved by the FDA in the 1980s are intraarticular, more than a decade before marketing of HA in the U.S. for human IA injection. However, it is noted that the product Legend can be used both intraarticularly and intravenously. Legend has been approved for IV use and has data from controlled trials to support the efficacy of the material.

Intravenous injection of Legend was shown to be as effective as IA injection, according to the *Freedom of Information* for Legend (NADA 140-883). The Legend injectable solution was a 1% injectable solution of NaHA manufactured by MOBAY. The 36 horses with experimentally induced carpitis (using Freund's complete adjuvant) were evenly divided into six treatment groups. Each group received one injection except Group 5, which received two injections:

1. Phosphate buffered saline, negative control (2 ml)
2. 5 mg Legend injectable solution (0.5 ml)
3. 20 mg Legend injectable solution (2 ml)
4. 40 mg Legend injectable solution (4 ml)
5. 20 mg repeated in two weeks (2–2 ml)
6. Positive control commercial product (Hylartin-V, 2 ml).

The evaluation of the effectiveness of HA for IV injection was based on clinical locomotion parameters (e.g., flexion, stride, lameness). "Clinical judgment of the investigator indicated that all but the placebo and 5 mg

TABLE 6.2

HA-Based Products for Race Horses Approved by the U.S. Food and Drug Administration

Product	Original Sponsor	Route of Administration	NADA Number (Approval Date)
Equron	Solvay Veterinary	IA: fetlock, carpal, hock	139-913 (7-10-86)
Hyalovet	Fort Dodge Laboratories	IA: fetlock, carpal	140-806 (5-31-88)
Hylartin V	Pharmacia & Upjohn	IA: fetlock, carpal, hock	112-048
Legend	MOBAY Corporation	IV and IA: fetlock, carpal	140-883 (9-12-91)
Synacid	Sterivet Laboratories	IA: fetlock, carpal	140-474 (6-1-88)
Hyvisc	MedChem Products	IA: fetlock, carpal, hock	122-578 (3-24-86)

IA: intraarticular, IV: intravenous.

Legend injectable solution doses were equally effective." Virtually complete alleviation of lameness occurred from the fourth to sixth week. Stride appeared mostly resolved by the time of the second week. Swelling resolved more slowly and less completely during the overall observation period.

It is interesting that related to the Legend product, a patent by Schultz et al. was assigned to Mobay Corporation in Pittsburgh, Pennsylvania (U.S. Patent 4,808,576). The title of the patent is *Remote Administration of Hyaluronic Acid to Mammals*.[100]

What could be the scientific rationale behind such phenomena? Kawcak et al. also found that as compared with placebos, horses treated with HA IV had better (lower) lameness scores and synovial membrane histologic scores.[56] In addition, the researchers observed in the HA-treated group a significantly lower concentration of total protein and prostaglandin E2 within synovial fluid 72 days after the surgery. It was suggested that IV-administered HA "appears to alleviate signs of lameness by interacting with synoviocytes, and by decreasing production and release of inflammatory mediators."[56]

It is important to note that systemic effect can come from IA injection, as well as IV injection. It is conceivable that a significant portion of the effect of HA locally injected may occur systemically. Intravenous injection is not indicated for human use, and because any viscous material brings the potential risk of clogging the blood vessel, such use is not likely to become reality in the future. Intramuscular use of HA might be a viable alternative, if proven safe and effective in clinical studies.

6.2.4.2   The Effect of Molecular Weight

The concept of viscosupplementation started from the scientific observation that synovial fluid in pathological conditions has significantly lower viscosity and lower MW of HA than what is found in healthy joints. This conceivably provided the rationale for using highly viscous material in expectation of a vastly improved performance. However, the rationale per se is not the proof of the concept, which can only come from experimental data. The key issue is whether there is strong evidence that higher-viscosity HA works better than lower-viscosity HA in its therapeutic performance related to the treatment of the symptoms of OA.

Some evidence from animal studies indicates that the ability of HA to relieve pain may be MW dependent. The degree of the pain relief resulting from HA injection, according to a cat study by Pozo et al., can be measured by the decrease of nerve responses from nociceptive afferent fibers in arthritic joints.[57] In 2004, Gomis et al. published a study on the effect of the MW of three HA preparations on articular nociceptor afferents in a rat model.[58] They found that

1. the lowest-MW Hyalgan did not reduce nerve impulse frequency of the evoked discharges in either normal or inflamed joints

2. the medium-range-MW HA product Orthovisc "had no effect in normal joints, but produced a transient frequency reduction of the evoked discharge in inflamed joints"

3. the highest-viscosity product Synvisc reduced the impulse discharge by 50% in both normal and inflamed joints during the period of 50 minutes to approximately 2 hours after injection

4. the nonviscous buffer control had no such effect

It appears in that study, the pattern of behavior of HA products was consistent with the viscosupplementation theory. The authors, however, cautioned against "assuming that the pain relief observed in human osteoarthritic knee joints is directly comparable" to one animal study. The joint pathology developed in the acute rat arthritis model is different from the pathophysiology of any painful human arthritis. Moreover, the animal studies were not systematic or well controlled. Many data points over time regarding the effect of pain relief postinjection were missing.

The effect of the MW of HA has been extensively investigated in numerous animal studies. Ghosh and Guidolin reviewed the data of at least eight HA preparations with MW ranging from 0.5 million (Bayer, Fidia) to 6.9 million (Pharmacia & Upjohn) dalton in approximately 25 animal studies including those of rabbit, beagle, mongrel dog, fox hound, and sheep. In those studies, OA was induced by such methods as anterior cruciate ligament (ACL) transection and menisectomy. Cartilage loss and gait analysis were among the outcomes for evaluation.[59] The review pointed out that the results from animal studies did not support the notion that "high MW-HA preparations were more biologically active than HA of lower MW." Instead, in large animal models of OA, the HA products (0.5×10^6 to 1.0×10^6 Da) were in general more effective in reducing inflammation in synovial fluid and inducing the restoration of its rheological properties than those HA products with MW greater than 2.3×10^6 Da. One of those examples mentioned in the review by Ghosh et al. was the study conducted by Asari et al.,[60] who found that the lower-MW preparation of HA (0.84×10^6 Da) had a better effect in reducing the inflammatory changes in ACL-deficient dogs than the high-MW preparation (2.3×10^6). Asari et al. used fluorescein-labeled HA and observed the enhanced penetration of the lower-MW HA through the extracellular matrix of the synovium. This was viewed as a viable explanation for the efficacy of lower-MW HA.[59]

Although the data from animal studies are more relevant to the product performance than the data from the *in vitro* studies, the real significance of MW of HA-based medical products in treating OA has to be borne out by the human clinical experience. Had there been sufficient evidence at a clinical level to support it, the purported mechanism based on local action would have been more widely accepted. Taking the intravenous injection of HA for the OA of racehorses as an example, the proof for systemic effect was not only based on the fact that it did relieve the symptom of lameness and pain but also on the fact that it did so in a dose-dependent manner. Similarly, MW of HA, if truly important to the treatment of OA pain, should have demonstrated a clear pattern of MW-dependent clinical performance. So far, the human clinical experience has pointed otherwise.

Efforts were made early on to gather clinical evidence in support of the importance of HA MW and viscosity. In fact Clinical Study 7 in the PMA submission of Synvisc was designed to explore the relationship of effectiveness of the product in relieving pain and its viscosity. Based on Dr. Schroeder's speech in the Orthopedic and Rehabilitation Devices Panel of the FDA, (November 20–21, 1996), the study, conducted in Germany, enrolled 132 patients, and four HA-containing formulations were tested. In the order of decreasing elastoviscosity, they were

1. Synvisc, a chemically crosslinked HA
2. Healon, a natural HA with a MW of 2 million
3. Artz, a natural HA with a MW of 750,000
4. degraded, nonelastoviscous Synvisc

The only outcome in favor of the contribution of elastoviscosity was the difference between Synvisc and Artz. In only one out of four pain measures, "subject evaluations of weight bearing pain at week 12," Synvisc-treated subjects showed a mean improvement of 38 ± 4 mm on VAS. This was a significantly greater improvement ($p = 0.03$) than that experienced by subjects treated with the 750,000-Da MW hyaluronan." However, in the same study, the least viscous formulation, the degraded, nonelastoviscous Synvisc, also demonstrated a 36-mm improvement for the weight-bearing pain, which is so close to the effect of Synvisc that no statistically significant difference between the two could be found. This is contrary to the expectation based on the theory of viscosupplementation. Furthermore, there was no statistically significant difference in any measures between group Synvisc and Healon. This difference was highly anticipated and badly needed to justify the development of Synvisc. In its comment on Study 7, the FDA concluded that "the study did not conclusively demonstrate the relationship of effectiveness and elastoviscosity." The PMA did not provide the complete data for Study 7. Later, Wobig et al. published[61] only the part of the data comparing the two arms of Synvisc and Artz. The propriety of publishing data selectively was questioned by Allard and O'Regan.[62] Recently, Karlsson et al. conducted a clinical trial of the high-MW Synvisc and low-MW Artz and compared their effect on weight-bearing pain. The trial showed no difference between the two treatments during the first 26 weeks.[63]

The data from Study 7, taken as they were presented, at least indicate that MW of HA is not critical to its clinical performance in the treatment of OA knees, once it is over a certain level (e.g., 750,000 Da), and that when MW exceeds a relatively high level such as 2 million dalton, it may not matter at all. A logical question follows. For the IA use of HA, what is the rationale for the extra effort and cost to chemically crosslink HA, along with the extra risks associated with it, when there are so many choices of high-MW HA products already available, such as Healon?

What would be the consequences in the treatment of OA of the knee, if the MW of HA were reduced to the level of its oligosaccharides? Because no such products have been developed and tested for human knee injection, there are no exact answers to that question. However, it is generally recognized that the biological properties of HA polymers, with MW greater than hundreds of thousands of daltons, may be vastly different from those of HA oligosaccharides with a MW of only a few thousand daltons. For example, Knudson et al.[64] found in an *in vitro* study (using articular cartilage tissue slices) that "HA oligosaccharides induced a dose-dependent state of chondrocytic chondrolysis." However, within a wide range of MW of HA polymer, it appears that the difference in its relative size of the HA molecule is not critical to its clinical performance. Contrary to the widespread belief that the higher the MW the better, there appears to be a loosely defined MW range within which HA may have the optimal therapeutic effect. Note the similar observation from animal studies, as Ghosh put it, "the most pharmacologically active HA preparations would be those that were small enough to gain access to the pericellular matrix but still large enough to elicit the required cellular response at its receptors and/or control the movement of important mediators to and from the cells."[59] In summary, the MW of HA does matter to a certain extent, but its influence on the utility of the HA product needs to be further investigated and properly defined.

6.2.4.3　Is HA Residence Time in the Knee Really Important?

Another truth-and-myth debate related to the mechanism of action is whether the efficacy of the IA HA in relieving the knee pain is mainly dependent on its residence time in the knee. The widely held belief is that the effect of HA should at minimum last as long as the material stays in the joint, if not much longer. This appears to be another rationale behind the crosslinking of HA, as crosslinking prolongs the residence time of the modified HA in the joint.

Lindqvist et al. in Uppsala, Sweden, investigated the elimination of the Restylane type of HA product (NASHA) from human knee joint.[65] In the study, six healthy human volunteers received a single injection (3 ml of 20 mg/ml) of ^{131}I-labeled NASHA. Radioactivity in the knee, blood, urine, and so on was measured. Magnetic resonance and gamma camera imaging were monitored. There appeared to be three distinct fractions of NASHA with elimination half-lives of 1.5 hours, 1.5 days, and 4 weeks. The authors suggest that the last phase of elimination represents the "true half-life" of NASHA in the joint and the "labeling method used causes minimal modification of hyaluronan."

Durolane for IA injection contains 3 ml of NASHA at a concentration of 20 mg/ml (Q-Med 2000 annual report: "NASHA contains 20 mg HA/ml and one injection with DUROLANE is 3 ml"), and it began sales in Europe in May 2001 (Q-Med 2001 annual report). Compare the half-life of the following HA preparations: HA Hyalgan in the knee, 17 hours (Fiorentini, R., *United States Food and Drug Administration Advisory Panel on Orthopaedic and Rehabilitation Devices*, November 21, 1996); hylan A component of Synvisc, 1.5 days; and hylan B component of Synvisc, 8.8 days (Berkowitz, D., *Proceedings of the United States Food and Drug Administration Advisory Panel on Orthopaedic and Rehabilitation Devices*, November 20, 1996). If the effect of three injections of Synvisc could last for up to 26 weeks, then it appeared reasonable to expect that Durolane, with three to four times the half-life of Synvisc, should remain effective for the same period of time even with one injection. That seemed to be the rationale of the product design of Durolane.

Q-Med stated in its 1999 annual report that Durolane "is expected to be able to be given as a single dose and still have at least the equivalent clinical effect of the competing products with their 3 to 5 injections. . . . Q-Med intends to document that Durolane eases pain and increases mobility in the affected joint for at least six months." The 2001 annual report of Q-Med stated that a major clinical study with 340 patients in the U.S., Sweden, and Canada was designed and conducted to confirm the above prediction. In March 2002, the results from the clinical study turned out to be a "disappointment" (Q-Med 2002 Annual Report on DUROLANE study). The study did not show "a statistically significant effect of treatment with Durolane compared with sodium chloride solution after 6 months if all clinics were taken into account" (Q-MED Press Release, April 23, 2002, Uppsala, Sweden). Similar to the cases of many other HA products for knee injection, when one looks hard at the data and performs "deeper analysis," somewhere and somehow a subgroup or two would appear as better than the others, even with statistical significance. The sponsor of the Durolane pointed to the discrepancy of patients' inclusion criteria — 130 patients with general OA did not do as well as the 216 patients with OA only in knees. After discussions with the FDA in October 2002, Q-Med announced in its 2003 annual report the decision to conduct a follow-up study on patients with "only osteoarthritis of the knee joint." The new planned study, however, would have a duration for only 6 weeks, instead of the previous study's duration of 6 months (Q-Med 2003 Report: "Recruitment of patients for a six-week study on DUROLANE is ongoing. Approximately 70 percent of the patients were recruited in January 2004. The study will form the basis of a supplementary application to the FDA").

6.3   Adhesion Prevention and Wound Healing

Scarring and fibrosis are a common occurrence following surgical procedures, such as intraperitoneal fibrous adhesions after abdominal surgeries, epidural fibrosis after spine surgery, and tendon adhesion after hand surgery. Strictures occur at anastomotic sites of bowel, adnexa, blood vessel, trachea, ureter, or bile duct; skin incisions heal with scars and sometimes with keloids; and transmission loss may occur after peripheral nerve injury. In abdominal and gynecological surgeries, bowel obstruction, abdominal pain, and infertility are associated with adhesions, with a major effect on the cost of health care.[66]

The HA-containing products approved in the U.S. for the prevention of postsurgical adhesion in certain OB-GYN and abdominal procedures (Table 5.3), and the HA-containing products approved for hemostatic dressing or stent in certain nasal and otologic surgeries (Table 5.4) have been summarized in Chapter 5. These two types of products are similar in the sense that they need to serve as a barrier between tissues not only during surgery but also for a minimum length of time after surgery. It is by design that all those products contain chemically modified HA, which in general extends the *in vivo* residence time of the HA material.

6.3.1   Design Rationale and Key Product Characteristics

6.3.1.1   *Mimicking Scarless Wound Healing of Fetus*

In adult, HA in the wound fluid peaks at 3 days and disappears at 7 days, whereas the tissue repair of fetus is characterized by the early appearance and prolonged maintenance of an extracellular matrix rich in HA. It is known that, unlike in adults, healing in fetus occurs rapidly in the absence of acute inflammation, and collagen is deposited in a highly organized, scarless manner. Although the detailed mechanism of fetal wound healing is not exactly clear, the correlation between scarless wound healing and the existence of large amount of HA is unequivocal. HA and its derivatives may have a role in the control of wound healing process: reepithelialization, inflammation, angiogenesis, contraction, and extracellular matrix synthesis, deposition, and remodeling.[67,68]

In fetal wound fluid, a glycoprotein with HA-stimulating activity was identified that is believed to be responsible for the prolonged presence of HA in fetal tissues.[68] Conversely, in adult wounds "the early deposition of HA occurs through the platelet plug and the fibrin clot. After the initial deposition of HA, hyaluronidase is produced, HA is removed, sulfated glycosaminoglycans then are deposited and finally collagen is laid down in a scar pattern."[67] The fetus is teaching us about wound healing, and the therapeutic goal is to apply those lessons to control the quality of healing.

6.3.1.2   *Minimum Duration of the Presence of a Barrier*

The summary of safety and effectiveness of Seprafilm (P950034) stated that "the mechanism by which barriers reduce adhesion formation involves physically separating damaged peritoneal surfaces from apposing surfaces during the early stages of wound repair."[69] Reperitonealization begins within 3 days after tissue damage. Once the reperitonealization is complete, the likelihood of adhesion formation between tissue surfaces is greatly reduced.

Using a mouse uterine horn model, Sawada et al.[70] conducted a study of Seprafilm, which demonstrated that the adhesions were formed within 3–5 days after the peritoneal injury, and reperitonealization was completed within 7 days. The fact that there was no difference in adhesion scores between day 7 and day 14 also indicated that adhesions were largely completed before day 7. It is therefore crucial for the injured surfaces to be separated with a barrier during the first 7 days postsurgery. It is likewise important that the peritoneal residence of the adjuvant should last as long as the period of adhesiogenesis. Seprafilm membrane became a hydrogel within 24–48 hours by absorbing water in the body fluid that is absorbed (not visible at the site) within 7 days and excreted in less than 28 days (*Summary of Safety and Effectiveness Data, P950034*).

6.3.1.3 HA Prevents Adhesion Better if Applied before Injury

HA is a component of the extracellular matrix of the peritoneal epithelium and is produced by peritoneal epithelial cells.[70] Urman et al.[71] demonstrated that HA prevents adhesion more effectively when it is applied before injury. Postoperative intraperitoneal adhesion formation was significantly reduced as compared to PBS and blank in a rat model, when diluted HA solutions (0.25 mg/ml and 0.4 mg/ml) were applied to precoat serosal surfaces before they are exposed to laser. However, there was no effect of HA solutions when applied at the end of the surgery.[71] This indicates the role of HA solution as a lubricant and antidesiccant in reducing trauma, but not as an effective barrier in preventing adhesion once the trauma has been inflicted. This provided the rationale of using a large volume of HA-containing solutions to "lavage" the whole exposed area of tissue surface before surgery to prevent the so-called de novo adhesion.

6.3.2 Products in Solid Forms

HA-based adhesion-prevention products indicated for abdominal and pelvic surgeries, like HA products as an aid for eye surgery or for injections into the arthritic knees, are classified as class III medical device. For market approval in the U.S., the sufficient evidence of effectiveness as well as safety supported by human clinical data is normally required. Johns reviewed the evidence-based prevention of adhesions post–gynecological surgery, and referred the evidence from well-designed, randomized, controlled trials as level I evidence.[72]

6.3.2.1 Seprafilm

The effectiveness of Seprafilm was evaluated mainly in two pivotal human clinical studies. The data were part of the PMA (P950034) submitted to the FDA for the approval of Seprafilm. These studies were also summarized in the package insert of Seprafilm, later also published in peer-reviewed journals.[73,74]

The first Seprafilm study enrolled 183 patients with ulcerative colitis and familial polyposis who underwent a special abdominal surgery that involves "colectomy followed by ileal pouch and anastomosis and formation of a temporary loop ileostomy." The Seprafilm membrane, averaging about 400 cm^2 per patient, was "applied directly on the omentum and bowel to separate tissues from the overlying abdominal wall and midline incision."[74] Several weeks later, another procedure called ileostomy closure was performed, during which the incidence, extent, and severity of adhesions to the middle incision were evaluated (Package Insert, Seprafilm Adhesion Barrier, http://www.genzyme.com, December 2004).

The main outcome measures in terms of the incidence, extent, and severity of postsurgical adhesions in the abdominal surgery for both Seprafilm treatment group and the control (no treatment) group were summarized in Table 6.3. In short, the differences between the groups favoring Seprafilm were statistically significant in all three measures. The magnitude of the difference could be appreciated even without statistical calculation. The incidence of adhesion for Seprafilm was only about 50% of the control.

The second pivotal clinical study of Seprafilm was to investigate its OB-GYN application. The study included 127 women undergoing myomectomy. The Seprafilm was applied to the anterior and posterior surfaces of the uterus following a myomectomy via laparotomy. Among Seprafilm-treated patients, the average amount of Seprafilm applied per uterus was 391 cm^2 (range 116–697 cm^2). This was equivalent to slightly greater than two sheets of Seprafilm (12.7 × 15.2 cm).[73] Postoperative adhesion to uterus was evaluated during a second-look laparoscopy performed 23 days later (package insert of Seprafilm).

The results of the main outcome measures for adhesions of the Seprafilm treatment groups and the control (no treatment) groups for OB-GYN surgeries

TABLE 6.3

Outcome Measures and Effectiveness Data of Seprafilm Clinical Trials for Premarket Approval — Abdominal Surgeries[a]

Clinical Outcome Measures Effectiveness Data[b]	Seprafilm	Control Nontreatment	P Value
Incidence	49%	94%	$P < .0001$
Presence of one or more adhesions to the area of the midline incision	(n = 42)	(n = 85)	
Extent	23%	63%	$P < .0001$
Percentage of midline incision Length with adhesion	(n = 85)	(n = 90)	
Severity	Grade 3:	Grade 3:	Overall adhesion in
1. Filmy thickness, avascular	15%	58%	Seprafilm group
2. Moderate thickness, limited vascularity	(n = 13)	(n = 52)	less severe than control ($P < .0001$)
3. Dense thickness, vascularized			

[a] "Restorative proctocolectomy and ileal J-pouch anastomosis with diverting ileostomy followed by second-stage laparoscopy for ileostomy closure and direct visual assessment of the peritoneal cavity."[74]

[b] Second look during subsequent ileostomy closure 6–12 weeks later, by laproscopy.[74]

were summarized in Table 6.4. Similar to the results for Seprafilm in abdominal surgeries, the differences between the groups favoring Seprafilm were statistically significant in all four measures (incidence, area, extent, and severity). Particularly important was that "the portion of patients with adnexal adhesions to the posterior uterus was reduced from 69% (n = 45) to 52% (n = 28) in patients with Seprafilm compared to control patients (p < 0.05)." In gynecology, the adnexa are the appendages of the uterus; namely, the ovaries, fallopian tubes, and ligaments that hold the uterus in place. Adnexal adhesions are a common cause for infertility.

The acceptance of using adhesion as surrogate end points for HA used in general surgery was not without controversy. Adhesion per se may not be considered as a disease, unless it affects the function of the tissues and organs involved. For myomectomy, the ultimate evidence of clinical utility would be the positive effect of the adhesion barrier on the fertility of the patients. However, such evidence is difficult to establish. For abdominal surgery, one of the more direct clinical end points would be the effect of the HA barrier on the incidence of small bowel obstruction.

Genzyme has started a postmarketing clinical study to further investigate the safety and effectiveness of Seprafilm for patients (the majority of patients

TABLE 6.4

Outcome Measures and Effectiveness Data of Seprafilm Clinical Trials for Premarket Approval — Myomectomy via Laparotomy

Clinical Outcome Measures, Effectiveness Data[a]	Seprafilm	Control — No Treatment	Between-Group Difference
Incidence:			
Number of abdominal-pelvic locations adherent to uterus[b]	4.98 (n = 49)	7.88 (n = 48)	$p < .0001$
Area:			
Area of uterus associated with adhesions	13.23 cm² (n = 65)	18.72 cm² (n = 54)	$p < .02$
Extent:	1.23	1.68	$p < .01$
0: no adhesions			
1: < 25% of location area			
2: 25–50% of location			
3: > 50% of location			
Severity:	1.94 (n = 54)	2.43 (n = 65)	$p < .01$
0: no adhesions			
1: filmy, avascular adhesions			
2: vascular or dense adhesions			
3: cohesive adhesions			

[a] Adhesion formation was evaluated during a second-look laparoscopy performed 23 days later.
[b] The 13 pelvic surfaces evaluated were anterior and posterior uterus, anterior and posterior cul-de-sac, right and left pelvic sidewall, right and left ovary, right and left tube, small and large bowel, and omentum.[73]

had inflammatory bowel disease) undergoing abdominopelvic surgeries. Between 1998 and 2000, 1791 patients were randomly assigned to the Seprafilm group or the nontreatment group.[75] The safety data were already reviewed in Chapter 5. The second part of the study was to examine the effectiveness of Seprafilm, including its reduction of small bowel obstruction. The 2–3-year follow-up data of the study would be reported after the completion of the study. In the meantime, Kudo et al. reported that "the incidence of early postoperative small bowel obstruction was significantly lower (P < 0.05) in the Seprafilm group (0/21) than in the control group (6/30)" after transabdominal aortic aneurysm surgery.[76]

In general, Seprafilm has demonstrated clear efficacy as compared to no treatment in certain abdominal and gynecological procedures. However, as the sponsor of Seprafilm acknowledged, "in both human and animal studies Seprafilm has not been completely effective in preventing postsurgical adhesion development. A number of factors may potentially reduce membrane effectiveness. These include membrane movement away from the application site, insufficient coverage of potential adhesiogenic sites due to improper membrane application" (*Summary of Safety and Effectiveness Data, Seprafilm Bioresorbable Membrane, P950034*). It is noted that in the recent large-scale clinical studies involving 1791 patients mentioned above,[75] the average number of films used per patient was 4.4, and as many as 10 were used for certain individual patients, as compared to slightly over two films used per patient in the two pivotal studies for PMA review mentioned earlier. The large increase in the amount of the adhesion-prevention material applied in this study conceivably was aiming at improving its ability to reduce adhesion in a wide scope of abdominal applications. However, using more material does not necessarily bring better results. In addition to the extra cost of material and adding excessive burden to the clearance, the persistent presence of a large amount of HA may actually hinder the critical healing process. The fresh anastomotic sites needs rapid healing, as otherwise anastomotic leaks would likely occur. Wrapping the suture or staple line with Seprafilm can cause overloading of the HA material at the fresh anastomotic sites, and therefore should be avoided.

Certain features of Seprafilm are not particularly user friendly for the intended purpose. Seprafilm is difficult to handle during surgical procedures. It is brittle when dry and sticky when moist and is therefore difficult for surgeons to make adjustments in applying it to the exact location during operations. It cannot be used in laproscopic surgery either. As a consequence, a more pliable version of Seprafilm II has been designed to overcome such shortcomings. In addition to carbodiimide-modified HA and CMC, a third component, glycerin, is added to Seprafilm II. The product is being marketed in Europe (http://www.genzyme.de).

6.3.2.2   MeroGel

MeroGel is composed of HYAFF, a chemically modified HA through esterification. The chemistry of HYAFF has been described in Chapter 2. MeroGel has two major indications, nasal and otologic.

MeroGel Nasal Dressing and Sinus Stent (Xomed), also known as Absorbant Intranasal Packing Material or Absorbant Intranasal Splint, is a white fibrous biomaterial. It can be used as a space occupying dressing in the nasal and sinus cavity, and it changes to viscous gel when in contact with body fluid. These characteristics of MeroGel allow it to function as a stent to separate mucosal surfaces in the nasal cavity and as a hemostatic material for minimal bleeding. The product was cleared for marketing by the FDA in 1999 (*Summary of Safety and Effectiveness, Xomed MeroGel Nasal Dressing and Sinus Stent, K982731*). Recently, MeroGel Injectable in the form of a viscous hydrogel has been introduced.

Miller et al. conducted a blinded, randomized, and controlled study of MeroGel, comparing it with Merocel, a nonabsorbable packing composed of synthetic polymeric material — hydroxylated polyvinyl acetal.[77] A total of 37 patients with chronic sinusitis underwent bilateral functional endoscopic sinus surgery, and MeroGel and Merocel were placed into left or right middle meatus of each patient in a randomized way. The rate of synechia by blinded evaluation at 8 weeks was considered as the primary study end point. It turned out that 8% of the MeroGel as well as 8% of the Merocel patients developed adhesion at 8 weeks, and it was obvious that there was no difference between the two nasal dressings.

Catalano et al. conducted a controlled study comparing MeroGel with Gelfilm (Pharmacia & Upjohn; made from denatured porcine collagen) as middle meatal stents for patients with chronic sinusitis after the surgery using minimally invasive sinus technique.[78] 100 patients were enrolled, with the right side inserted with MeroGel and the left side with Gelfilm. There was a significant difference ($P = .04$) in synechia between the MeroGel (none) and Gelfilm (4), and a significant difference ($P = .0004$) in stent retention time between MeroGel (7.9 days) and Gelfilm (5.6 days). It is noted that the overall rate of synechia with stent in this study (0–4%) was quite low. According to Gaskins, who reviewed the data of 970 endoscopic nasal surgeries,[79] the average synechia rate of various stents was 6.7%. The lower rate in Catalano's study could be attributed to the minimally invasive sinus technique used versus other endoscopic techniques. Gaskins also found in the same study that without the stent, the average synechia rate was 10%. The limited information available appears to support the clinical utility of stenting material in sinus surgeries. However, no general conclusions with regard to the efficacy of HA-based materials relative to other materials can be drawn. Another interesting characteristic of MeroGel is that the material may be osteogenic, as indicated by a mice study conducted by Jacob et al.[80]

To investigate the otologic application of HYAFF type of biomaterial [the Xomed Merogel Otologic Pack is composed of HYAFF; 510(k) K001148], Martini et al. conducted a clinical study to test its effect in stimulating the healing process in patients undergoing tympanomastoid surgery.[81] The reepithelialization of open cavities and tympanic grafts is somewhat unpredictable. It was hoped that HA-based biomembrane would reduce such variability. In the study, 60 patients with chronic cholesteatomatous otitis media were enrolled (a cholesteatoma is a skin growth that usually occurs in the middle ear (ME) behind the eardrum, commonly caused by repeated infection. The choleste-atoma can increase in size over time and destroy the surrounding delicate hearing bones of the middle ear). The treatment group received HYAFF 11 (benzylester of HA), whereas the control group was given no biomembrane. The outcome measures, assessed at several time points between 3 and 25 weeks, included patients' assessment of quality of healing (pain, secretion), otomicroscopy (otorrhea, granulating tissue, new eardrum perforation and vascularization, reepithelialization), and audiometry. The patients in the biomembrane group had statistically significantly better scores than controls for all variables considered, except audiometry. The results showed that HA-based biomembrane "improves the healing of the mastoid cavity following open tympanoplasty. In particular, re-epithelialization was significantly faster after the application of the biomembrane. Moreover, the device neither caused any adverse events nor showed signs of ototoxicity" (p. 468).

It is not clear whether the chemically modified HA actually performs better than other types of biomaterials, such as gelatin-based material or even the native HA, under the condition of human ear surgeries. Extra benefit from using HA derivatives, after all, was generally expected and included in the design rationale of such products. Li et al.[82] compared MeroGel with absorb-able gelatin sponge in a study with guinea pigs. They found that on post-operative day 5, all ears with ME packing had hearing losses in the frequency range of 500–4000 Hz. The recovery of hearing acuity on postoperative week 6, however, was significantly better in MeroGel-treated guinea pigs than in the animals treated with the gelatin sponge.

Concurrently, it was observed that 20% of the absorbable gelatin sponge packing material remained in the ME cavities, and there was new bone formation, whereas for the MeroGel group, less packing material was found in the ME, and there was no formation of new bone. The results indicated that the MeroGel actually performed better in terms of causing less hearing loss in the first 6 weeks, and this coincided with a shorter residence time in the ear than the gelatin sponge. One may wonder what the results might have been if materials with a residence time shorter than MeroGel, such as high-MW native HA, had been used as a control?

Although native, unmodified HA for otologic use has not gained market-ing approval in the U.S., its effect on the healing of tympanic perforation as demonstrated in human clinical trials has been known since the 1980s.[83–86] A Healon-type of HA product was used in human clinical studies that demonstrated reduced size of the perforation and accelerated healing of the tympanic membrane. Nevertheless, information of direct comparison between native HA and chemically modified HA is not available.

6.3.3   Products in Liquid Forms

Adhesions at the sites outside the direct area of surgical trauma are referred to as *de novo* adhesions. They may be caused by, for example, surgical gauze or the powder on the surgeons' gloves. Because *de novo* adhesions may occur in unpredictable locations away from the surgical site, it was thought that a product in the form of a liquid, at least in concept, should be better suited to cover those areas. The *de novo* adhesion may occur in many anatomical sites within a large area; therefore, a large volume of the antiadhesion liquid is needed to cover all the areas that may have been exposed during abdominal–pelvic surgeries. Sepracoat and Intergel are two of such examples.

6.3.3.1   *Sepracoat*

Sepracoat is a 0.4% HA solution developed by Genzyme. In a human clinical study using Sepracoat during gynecological laparotomy, over 200 patients were enrolled, and the control group was given phosphate buffer solution. The blinding was possible because of the low viscosity of the Sepracoat resulting from its low HA concentration (0.4%). In total, 23 anatomical sites were identified, where the incidence of adhesion was observed by second-look laparoscopy. The results of the study were presented to the FDA and reviewed by its General Plastic Surgery Devices Panel during the panel's meeting on May 5, 1997.

The clinical outcome measures as primary end points in that study were defined as the proportion of number of sites with adhesion over the total number of available sites for *de novo* adhesion, and the percentage of patients who developed at least one adhesion postsurgically.

The results pertaining to the first end point were 29.5% of adhesion in control group and 23.6% in Sepracoat group. According to the interpretation by the sponsor, this constituted about a 20% difference and was significant. The FDA statistical reviewer (Dr. Lin, a speaker at the General Plastic Surgery Devices Panel), however, pointed out that the truly meaningful data were the average number of locations with observed de novo adhesions, which were 3.83 for Sepracoat and 4.53 for the control (intent-to-treat population). The real difference between the two groups was only 0.7, less than one adhesion site out of a total 23. In the meantime, with regard to the second end point, the Sepracoat group had approximately 88.8% patients with at least one *de novo* adhesion of available sites, whereas the patients in the control group with at least one *de novo* adhesion were 95.4%.

It was quite obvious why the panel did not recommend the approval of Sepracoat. Intuitively, the differences mentioned above just looked too small to convince the panel of its clinical utility. Historically, in the Sepracoat study protocol submitted for FDA review (also described in the same panel minutes), the sample size was based on the following estimation and anticipation for incidence of de novo adhesions at second look; that is, "approximately 30% in the control, and the ability of the Sepracoat to reduce the *de novo* adhesions to approximately 10%. So that is an absolute difference of 20 percent. That was in the protocol determined to be clinically significant." The actual results coming out of the study were not even close (the panel voted unanimously that Sepracoat was nonapprovable.).

Certain tactics are often seen used by the sponsors or performers of the studies during and after the regulatory review process in response to the unfavorable opinions of the agency and its advisory panels. One such tactic, justified or not so justified, is to run analysis retrospectively. In the case of Sepracoat, incidence of adhesion was a predetermined primary end point, but extent and severity were not. After the panel's negative recommendation, however, more analyses on the extent and severity were performed or brought to highlight to show an improved difference.[87] Another tactic was to be "creative" in presenting numbers to give different impressions. For example, in the case of the first primary end point of Sepracoat (proportion of number of sites with adhesion over the total number of available sites for *de novo* adhesion) the simple difference between 29.5% and 23.6% was 5.9%, an intuitively small number. However, when that 5.9% was divided by 29.5%, a much more impressive relative difference of 20% was derived. Note the same approach was not taken to argue for the results of the second primary end point (percentage of patients who developed at least one adhesion postsurgically), which were 88.8% for Sepracoat versus 95.4% for control, with a "simple difference" of 6.6%. It is obvious in that case, 6.6% divided by 95.4% would have afforded only a likewise small number 6.9%, which certainly would not have helped with the case. But think again — would 88% versus 95% of patients with at least one incidence of adhesion be equivalent to 12% versus 5% of patients without adhesions?[87] After all, who can dispute that 12% is more than 100% better than 5%?

Whatever the controversy about the effect of a "lavage" HA solution may have on de novo adhesion, one indisputable fact is that a diluted native HA solution does not work as effectively as its solid barrier counterparts. If one believes in the mechanism of action is physical barrier, then the lack of efficacy of a lavage method should not come as a surprise. A 0.4% HA solution like Sepracoat contains over 99.5% of the water, and if this can be considered a barrier, it is certainly one with countless holes.

6.3.3.2   Intergel

Intergel is a dilute (0.5% w/v) ferric hyaluronate solution adjusted to isotonicity with sodium chloride. Unlike Sepracoat, which has to be added every 30 minutes during surgery, Intergel is an "intraperitoneal instillate at the completion of the surgery." This ionically crosslinked HA can also be efficiently metabolized by the tissues responsible for endogenous HA clearance. The residence time of Intergel following intraperitoneal administration appears to be a slightly longer than the native HA. The elimination half time of Intergel solution in humans is expected to be approximately 51 hours (*Summary of Safety and Effectiveness Data, Gynecare Intergel, P990015*).

The pilot study of Intergel enrolled 23 patients undergoing peritoneal cavity surgery by laparotomy. Thirteen patients received Intergel, and 10 received lactated Ringer's solution. The study was described in PMA 990015 and reported by Thornton et al. in a journal article as well.[88] The pivotal study in peritoneal cavity surgery submitted to the FDA was a randomized,

TABLE 6.5

American Fertility Society Score of Severity
and Extent for Each Adhesion Site

American Fertility Society Score per Site	Severity	Extent
0	None	None
1	Mild	Localized
2	Mild	Moderate
4	Mild	Extensive
4	Severe	Localized
8	Severe	Moderate
16	Severe	Extensive

281-patient study. One hundred forty-three patients received 300 cc Intergel, and 138 patients received the same volume of Lactated Ringer's at the end of a peritoneal cavity laparotomy. The procedures were clean in the sense that there were no potentially infectious processes involved, such as gastrointestinal or genitourinary tract breach. The study was also described in PMA 990015, and later in an article by Johns et al.[89]

Similar to the outcome measure in Sepracoat study, the Intergel study looked at the incidence, extent, and severity of the de novo adhesions. The difference was that Intergel used the unique scoring systems from the American Fertility Society (AFS).[90] The AFS scoring system evaluates 16 anatomical sites, and the modified AFS (mAFS) scoring system evaluates 24 anatomical sites (listed in Table 3 of PMA 990015; they include peritoneum, bowels, uterus, ovary, tube, and so on). The latter was used as the measure for primary effectiveness results in the Intergel pivotal study. Both systems integrate the extent (none, localized, moderate, and extensive) and severity (none, mild, severe) of each site into a 0–16 scoring range composite score, as tabulated in Table 6.5.

It is obvious that the AFS scoring system is heavily weighted on the degree of severity and extent. As a consequence, a decrease of few serious adhesions in the whole study would have a significant effect on the outcome. As illustrated in Table 6.6, Intergel was found to have statistically significant

TABLE 6.6

Primary and Secondary Effectiveness Measures of Intergel

Effectiveness Measures	Intergel	Control	P Value
Primary measures, mAFC (0–16)	1.36	2.32	0.002*
Secondary effectiveness measures			
Incidence of adhesion (0–24)	6.56	7.63	0.096
Extent of adhesion (0–3)	0.47	0.63	0.019*
Severity of adhesion (0–3)	0.52	0.73	0.007*

*Statistical significance.

effect in reducing pelvic adhesions in the pivotal study, in which such AFS scoring system was adopted as a primary outcome measure. In the group of secondary measures, in which the incidence, extent, and severity were analyzed separately, the between-group significance was found only in the extent and severity scores, but not in the incidence score. All baseline data were comparable.

The question remained as to whether the 1.07 difference of the incidence of adhesion out of 24 was clinically meaningful. The General Plastic Surgery Devices Panel of the FDA was particularly interested in how treatment affected the proportion of patients with moderate/severe adhesions at adnexa (the ovaries, uterine tubes, and ligaments of the uterus). The panel was concerned about the uncertain value of AFS scores at second look in predicting hard outcomes of interest (pain, fertility, bowel obstruction). Nevertheless, the panel accepted that the AFS score "probably was the best measure, given the constraints of what is feasible, for a prospective clinical trial. . . . The Panel stated that Intergel showed a statistically significant benefit over the lactated Ringer's solution control, and that this evidence was sufficient for approval." There was obvious reservation within the panel, who noted the modest clinical benefit of Intergel. The recent voluntary withdraw of Intergel from the market as a result of safety concerns about its off-label use was mentioned in Chapter 5.

6.3.3.3   HylaSine and Sepragel Sinus

HylaSine was made of hylan B, a crosslinked HA by divinylsulfone. The HylaSine gel was indicated for nose and ear surgeries to prevent adhesion and stop bleeding. The product was cleared for marketing in the U. S. through a 510(k) review process, and no human clinical study was required. The public information through regulatory approval documents with regard to its clinical performance and safety has been scarce. The product Sepragel Sinus by Genzyme has the same active ingredient hylan B, and the same indication as a nasal dressing and sinus stent, as well. Sepragel Sinus is registered as a class I medical device with the FDA (registration number 840.4780). Kimmelman et al. conducted a human clinical study to investigate how Sepragel Sinus would perform as a postsurgical dressing for endoscopic sinus surgery to facilitate healing and prevent scarring and stenosis.[91] Ten patients were recruited to undergo bilateral endoscopic ethmoidectomy (a surgical procedure during which ethmoid sinus cells are opened to treat infection or sinus obstruction that has led to chronic sinus problems), and Sepragel Sinus was used to fill a randomly selected right or left ethmoidectomy cavity. Outcome measures included synechia, middle meatal stenosis, mucosal status, mucosal regeneration, pain, and congestion. "Sepragel Sinus significantly improved all outcome measures by week 2 and remained statistically significant for reduction of synechia and stenosis." No information about direct comparison between crosslinked HA and native HA is available.

6.3.4 Challenge in Adhesion Prevention Products

It is important to note that HA-based adhesion-prevention products were designed and developed with the reasonable expectation that they would bring greater clinical benefit as compared to the products already on the market, such as Interceed and Gore-Tex Surgical Membrane. Both products are barrier-type products. Interceed (TC7; Ethicon, Inc.) is made of oxidized regenerated cellulose — a meshlike product designed to be placed over or between injured surfaces in both laparotomy and laproscopy. More than a dozen clinical trials were conducted but only demonstrated effectiveness in very limited situations, such as in the absence of blood. Gore-Tex Surgical Membrane (W. L. Gore and Associates), as a nonabsorbable barrier (expanded polytetrafluoroethylene),[92,93] must be sewn in place and is usually removed during a second surgical procedure. It is difficult to apply in laproscopic surgery.

The goal of developing an HA-based product is at best partially fulfilled. The function of HA as an adhesion-prevention material is not affected by the presence of blood, nor does it need another surgery to remove the HA barrier. They are, under the exact, indicated condition, generally safe. However, their effectiveness is only marginal, and the risk of improper use could be very serious.

The problems in the lack of efficacy and the susceptibility to infection in the use of some of the HA-based lavage-type products appeared to be a continuation of the problems of their predecessors. As an intraperitoneally administered liquid product, Hyskon (Medisan Pharmaceuticals) contains 32% Dextran 70. The product decreases adhesion by hydroflotation and affecting the clotting cascade. However, two of the four randomized clinical trials failed to show efficacy. In addition, allergic reactions have been reported. These combined, according to Johns, have "virtually eliminated" the use of Hyskon in gynecological surgery.[72]

To improve the design of HA-based adhesion-prevention products, a fundamental and broader concern needs to be addressed — that "a reduction in post-operative adhesions does not necessarily produce a better clinical outcome."[72] An example was the "leakage" problems encountered when using Seprafilm in anastomosis, mentioned in Chapter 5. There appears to be an intrinsic conflict between a material's ability to prevent adhesions and its property not to interfere with the healing process. Sometimes, a large amount of material is needed to prevent adhesion, but excessive HA presence can slow down the healing process. Therefore, finding a balance between the two and building that into the product design could be a real challenge.

The challenge to the adhesion prevention material also in part comes from the improvement in surgical techniques. The progress in microsurgery, precise hemostasis, constant irrigation, small and nonreactive sutures, minimal use of electrosurgery, and so on has reduced the magnitude of the problems of post-surgical adhesions. In the meantime, it is also more difficult to prove the effect of an adhesion-prevention material in a surgery that is minimally invasive.

In view of the effectiveness, safety, handling characteristics, and cost in their totality, the benefit of HA-based products developed so far for adhesion prevention can hardly be considered satisfactory. In complicated surgical situations involving large organ area, such as abdominal and pelvic surgeries, it may be too simplistic to expect that just by inserting a piece of HA material or instilling a bolus of HA gel into the body under surgery all adhesion problems would be resolved without unintended consequences. Much is still to be desired in the product design, which needs to be revisited in light of the clinical experience in the last decade.

6.4   Tissue Augmentation

HA-based products also found their way into the market of cosmetic tissue augmentation. Crosslinked HA products Restylane (Q-Med) and Hylaform (Genzyme) have been approved by the FDA for the treatment of facial wrinkles (Table 1.1). The characteristics and indications for these two products were summarized in Table 5.5. The clinical trials of Restylane and Hylaform used positive controls. This is different from the clinical studies conducted for other major HA indications (HA for the ophthalmic, abdominal/pelvic surgeries and for injections in OA knees), in which negative controls (or purported negative controls) were used. The positive control in the pivotal studies for Restylane and Hylaform was the collagen-based product Zyplast, which is a dermal collagen implant for aesthetic use and is considered the golden standard. Zyplast was approved by the FDA as a class III medical device (P800022) in 1980, when a controlled, rigorous clinical study to demonstrate effectiveness was not formally required. The acceptance of the collagen implant as a positive control in the studies for Restylane and Hylaform was based on more than 20 years of clinical experience with Zyplast. The FDA allowed the design of the studies of Restylane and Hylaform to be based on their noninferior comparison of effectiveness with the Zyplast.

6.4.1   Restylane

Restylane is composed of 20 mg/ml of NASHA, an epoxide crosslinked HA. According to the *Summary Lead Review Memo, Restylane Injectable Gel, P020023*, the pivotal study of Restylane for the Correction of Nasolabial Folds was a randomized, evaluator-blind, multicenter study that enrolled a total of 138 patients in the U.S. Restylane was injected on one side of the face, and Zyplast, a bovine collagen product, on the other. The response of the initial treatment was evaluated after 2 weeks, and touch-up procedures may be performed and repeated until optimal response — an established baseline — was achieved.

   The effect of Restylane, as well as the control, on the visual severity of the nasolabial folds was assessed by an evaluating investigator at 6 months

TABLE 6.7

Wrinkle Severity Rating Scale Score

SRS score	Description of SRS Scores
1. Absent	No visible fold; continuous line
2. Mild	Shallow but visible fold with slight indentation; minor facial feature
3. Moderate	Moderately deep fold; clear facial feature visible at normal appearance but not when stretched; excellent correction expected
4. Severe	Very long and deep; prominent facial feature; less than 2 mm visible fold when stretched
5. Extreme	Extremely deep and long folds; 2–4 mm visible v-shaped fold when stretched; detrimental to appearance; unlikely to have satisfactory correction with injectable implant alone

post-baseline. The evaluation parameter was a 5-point Wrinkle Severity Rating Scale (SRS) Score, as summarized in Table 6.7.

This scoring system was based on photo review. A difference in SRS equal to 1 is considered to be clinically significant according to a prestudy validation of the SRS scale. The sponsor argued for the claim of the superiority of Restylane over the control on the basis that the majority of patients receiving Restylane succeeded in achieving a 1-point improvement over the control in the SRS from baseline at 6 months. However, the FDA pointed out that in "the entire cohort of Restylane versus Control. . . a full one-point improvement could not be achieved in the aggregate case (D = 0.58)." To put in perspective of the magnitude of the effect of the treatment, bear in mind that the SRS scores of both groups were approximately the same at about 3.30 pretreatment and about 1.80 at baseline (1.5-unit improvement). The results from the measure of SRS score after 6 months were 2.36 for Restylane and 2.94 for Zyplast, which reflect the remaining improvement of Restylane (0.93) and the control Zyplast (0.36). Another result to support the beneficial effect of Restylane was that among the 137 patients (with baseline data), the SRS score of Restylane side was lower (better) for 88 patients, equal for 44 patients, and higher (worse) for only 13 patients (*Clinical Review of Restylane, Freedom of Information, P020023*). In summary, the numbers from the clinical studies appeared to favor the Restylane, but the question was whether it had reached a meaningful degree. The panel voted approval for Restylane after first voting to remove a superiority claim for Restylane (General and Plastic Surgery Devices Panel of the Medical Devices Advisory Committee, November 21, 2003).

6.4.2 Deflux

Another FDA-approved Q-Med product containing crosslinked HA for tissue augmentation is Deflux. Unlike Restylane, however, Deflux is not for cosmetic use, but for the treatment of vesicoureteral reflux (VUR) of children, an abnormal condition in which urine flows backward from the bladder to the kidneys, causing repeated, severe urinary tract infections. As a bulking agent, Deflux is injected submucosally in the urinary bladder in close proximity to the ureteral orifice.

There are two major ingredients in the formulation of Deflux, a crosslinked dextran ("dextranomer" 50 mg/ml) and an epoxide crosslinked HA (17 mg/ml) similar to Restylane. The dextranomer microspheres have an average diameter of approximately 130 μ. In the clinical and preclinical studies submitted to the FDA, it was found that the dextranomer microspheres remained in the tissue for at least 2 years postimplantation. The microspheres did not migrate and were gradually surrounded by host connective tissue. It appears that it is the crosslinked dextran that plays a major role in the durability of the tissue augmentation agent for treating VUR, whereas the crosslinked HA was referred to in the product label as a "carrier gel" (*Summary of Safety and Effectiveness Data, Deflux Injectable Gel, P000029*).

The utility of Deflux in the endoscopic treatment of VUR has been reported in medical journals.[94–97] Oswald et al. concluded: "Deflux, a new biocompatible, biodegradable substance, seems to be an alternative modality to other bulking agents for treating vesicoureteral reflux in children, with acceptable short-term and 1-year results."[97] Chertin and Puri commented, "in terms of effectiveness and long-term successful results, polytetrafluoroethylene is still the most reliable injectable material for the endoscopic treatment of VUR."[96]

6.4.3   Hylaform

Hylaform and Restylane were reviewed by the General and Plastic Surgery Devices Panel of the FDA in the same meeting, held on November 21, 2003. The formal notice of approval for Restylane was given on December 12, 2003, and for Hylaform on April 22, 2004. Hylan B, the active ingredient of Hylaform, however, was not a new chemical entity but was the same chemically as one of the components of Synvisc, approved by FDA 8 years earlier for injection to the knee of OA. Hylaform is a clear, transparent crosslinked HA gel with a concentration of 4.5–6.5 mg/ml, an osmolality of 290–330 mOsm, and a median particle size of approximately 500 μm (Table 5.5).

In the pivotal U.S. study at 10 dermatology centers, a total of 261 patients were enrolled, in which 133 subjects received Hylaform and 128 received Zyplast. The effect of the gels was assessed by Independent Panel Review photographic evaluation at 12–14 weeks. The wrinkle assessment score used was based on a validated, 6-point Genzyme Grading Scale that can be summarized as (0) no wrinkle, (1) just perceptible wrinkle, (2) shallow wrinkle, (3) moderately deep wrinkle, (4) deep wrinkle with well-defined edges, and (5) very deep wrinkle, redundant fold. A change of 1 point was considered as clinically significant. However, 12 weeks after treatment, the scores of Hylaform (2.3) and Zyplast (2.2) were virtually tied.

The effect of the dermal filler was measured "as the proportion of Hylaform treated nasolabial folds which returned to baseline scores at 12 weeks after last treatment. . . . Of the 243 total Hylaform treated folds, 178 (73.3%) returned to their baseline values. At 2 weeks the proportion was only 38.2%" (*Clinical Summary of Hylaform Viscoelastic Gel, P030032*).

Hylaform Plus was approved as a PMA supplement (P030032/S001) to the PMA of Hylaform (P030032). The formulation, manufacturing process, container, and packaging configuration are the same as Hylaform. "The only differences between Hylaform and Hylaform Plus include a slightly higher median particle size (700 microns vs. 500 microns) and a larger needle gauge (27 G vs. 30G)." This difference did not bring about the anticipated improvement in clinical performance. In the clinical trial comparing Hylaform and Hylaform Plus (96 patients each), the two products were found to be comparable for their clinical performance in the correction of nasolabial folds. Interestingly, the difference was between the methods of evaluation. Using independent review of photographs, patients of both groups returned to baseline at 12 weeks. When the assessment was made on the direct observation of the wrinkles by investigators or patients (as secondary endpoints of the Hylaform study), the effect of Hylaform or Hylaform Plus remained to be detectable at week 12 (*Summary of Safety and Effectiveness Data of Hylaform Plus, P030032/S001*).

6.5 Dental and Bone Application

In the area of bone repair, the role of HA as a carrier of demineralized bone powder was tested in a mouse model. Both the flowable paste formulation and the malleable putty formulations exhibited a positive effect on active bone formation when applied to the bony defects. No untoward inflammatory response was observed.[98] DBX®-Putty (MTF — Musculoskeletal Transplant Foundation), a product composed of Demineralized Bone Matrix (allograft) and HA, is currently available for sale in the U.S. (DBX; http://www. dentsplyfc.com). The product did not require FDA approval because it belonged to the category of allograft tissues, and HA was considered as a carrier. The HA component obviously contributes to the physical integrity and handling property and was claimed to make the osteoinductive DBX more effective than a low-MW carrier such as glycerol.

PepGen P-15™ Putty (Dentsply Friadent Ceramed), a bone graft for dental application, is composed of calcium phosphate mineral and a peptide-mimicking collagen (both osteogenic) and the carrier sodium hyaluronate. Similar to the case of DBX, HA markedly improved the handling characteristics by forming an easily moldable putty, which facilitates the application of the bone graft material to the site. PepGen P-15 Putty entered the U.S. market in 2004 via a supplement to the original PMA of PepGen P-15 (P990033/S005). PepGen P-15 does not contain HA carrier.

Similarly, HYAFF, a benzylester of HA, can be used as a scaffold for autologous fibroblast cell growth in gingival augmentation. The HA-containing graft demonstrated excellent compatibility with the fibroblast and even promoted the tissue growth.[99] HA benzylester is also a component of Hyaloss (http://www.metahosp.it; Hyaloss Matrix, CE 0459, Hyaloss Matrix brochure of META Advanced Medical Technology), an adjuvant in the surgical application of bone grafts treating dental defects. On contact with the patient blood or saline, the fibers turn into a gel that facilitates the application of bone fragments. The product is currently marketed outside the U.S.

References

1. Lang, E., Mark, D., Miller, F. A., Miller, D., and Wik, O., Shear flow characteristics of sodium hyaluronate. Relationship to performance in anterior segment surgery, *Arch. Ophthalmol.*, 102, 1079–1082 (1984).
2. Bothner, H. and Wik, O., Rheology of hyaluronate, *Acta Otolaryngol. Suppl.*, 442, 25–30 (1987).
3. Dick, H. B., Krummenauer, F., Augustin, A. J., Pakula, T., and Pfeiffer, N., Healon5 viscoadaptive formulation: comparison to Healon and Healon GV, *J. Cataract. Refract. Surg.*, 27, 320–326 (2001).
4. Hammer, M. E. and Burch, T. G., Viscous corneal protection by sodium hyaluronate, chondroitin sulfate, and methylcellulose, *Invest. Ophthalmol. Vis. Sci.*, 25, 1329–1332 (1984).
5. Glasser, D. B., Matsuda, M., and Edelhauser, H. F., A comparison of the efficacy and toxicity of and intraocular pressure response to viscous solutions in the anterior chamber, *Arch. Opththalmol.*, 104, 1819–1824 (1986).
6. Glasser, D. B., Katz, H. R., Boyd, J. E., Langdon, J. D., Shobe, S. L., and Peiffer, R. L., Protective effects of viscous solutions in phacoemulsification and traumatic lens implantation, *Arch. Opththalmol.*, 107, 1047–1051 (1989).
7. McDermott, M. L., Hazlett, L. D., Barrett, R. P., and Lambert, R. J., Viscoelastic adherence to corneal endothelium following phacoemulsification, *J. Cataract Refract. Surg.*, 24, 678–683 (1998).
8. Schubert, H. D., Denlinger, J. L., and Balazs, E. A., Exogenous Na-hyaluronate in the anterior chamber of the owl monkey and its effect on the intraocular pressure, *Exp. Eye Res.*, 39, 137–152 (1984).
9. Fry, L. L., Postoperative intraocular pressure rises: a comparison of Healon, Amvisc, and Viscoat, *J. Cataract Surg.*, 15, 415–420 (1989).
10. Mac Rae, S. M., Edelhauser, H. F., Hyndiuk, R. A., Burd, E. M., and Schultz, R. O., The effects of sodium hyaluronate, chondroitin sulfate, and methylcellulose on the corneal endotheliam and intraocular pressure, *Am. J. Ophthalmol.*, 95, 332–341 (1983).
11. Dick, H. B., Augustin, A. J., and Pfeiffer, N., Osmolality of various viscoelastic substances: comparative study, *J. Cataract Refract. Surg.*, 26, 1242–1246 (2000).
12. Poyer, J. F., Chan, K. Y., and Arshinoff, S. A., Quantitative method to determine the cohesion of viscoelastic agents by dynamic aspiration, *J. Cataract Refract. Surg.*, 24, 1130–1135 (1998).

13. Poyer, J. F., Chan, K. Y., and Arshinoff, S. A., New method to measure the retention of viscoelastic agents on a rabbit corneal endothelial cell line after irrigation and aspiration, *J. Cataract Refract. Surg.*, 24, 84–90 (1998).
14. Arshinoff, S. A., Dispersive and cohesive viscoelastic materials in phacoemulsification revisited 1998, *Ophthal. Pract.*, 16, 24–32 (1998).
15. Arshinoff, S. A. and Wong, E., Understanding, retaining, and removing dispersive and pseudodispersive ophthalmic viscosurgical devices, *J. Cataract Refract. Surg.*, 29, 2318–2323 (2003).
16. Hutz, W. W., Eckhardt, H. B., and Kohnen, T., Comparison of viscoelastic substances used on phacoemulsification, *J. Cataract Refract. Surg.*, 22, 955–959 (1996).
17. Arshinoff, S. A., Dispersive-cohesive viscoelastic soft shell technique, *J. Cataract Refract. Surg.*, 25, 167–173 (1999).
18. Miyata, K., Nagamoto, T., Maruoka, S., Tanabe, T., Nakahara, M., and Amano, S., Efficacy and safety of the soft-shell technique in cases with a hard lens nucleus, *J. Cataract Refract. Surg.*, 28, 1546–1550 (2002).
19. Behndig, A. and Lundberg, B., Transient corneal edema after phacoemulsification: comparison of 3 viscoelastic regimens, *J. Cataract Refract. Surg.*, 28, 1551–1556 (2002).
20. Arshinoff, S. A., Using BSS with viscoadaptives in the ultimate soft-shell technique, *J. Cataract Refract Surg.*, 28, 1509–1514 (2002).
21. Zetterstrom, C., Wejde, G., and Taube, M., Healon5: comparison of 2 removal techniques, *J. Cataract Refract. Surg.*, 28, 1561–1564 (2002).
22. Arshinoff, S. A., in *Rock 'n' roll removal of Healon GV, 7th Annual National Ophthalmic Speakers Program,* Medicopea, Quebec, 1997, pp. 29–30.
23. Tetz, M. R. and Holzer, M. P., Two-compartment technique to remove ophthalmic viscosurgical devices, *J. Cataract Refract. Surg.*, 26, 641–643 (2000).
24. Assia, E. I., Apple, D. J., and Lim, E. S., Removal of viscoelastic materials after experimental cataract surgery *in vitro, J. Cataract Refract. Surg.*, 18, 3–6 (1992).
25. Mamalis, N., OVDs viscosurgical, viscoelastic, and viscoadaptive. What does this mean? *J. Cataract Refract. Surg.*, 28, 1497–1498 (2002).
26. Kiss B., Findl, O., Menapace, R., Petternel, V., Wirtitsch, M., Lorang, T., Gengler, M., and Drexler, W., Corneal endothelial cell protection with a dispersive viscoelastic material and an irrigating solution during phacoemulsification: low-cost versus expensive combination, *J. Cataract Refract. Surg.*, 29, 733–740 (2003).
27. Soman, N. and Banerjee, R., Artificial vitreous replacement, *Bio-Med. Mater. Eng.*, 13, 59–74 (2003).
28. Balazs, E. A. and Sweeney, D. B., in *New and Controversial Aspects of Retinal Detachment*, McPherson, A., Ed., Harper Row, New York, 1968, pp. 371–376.
29. Pruett, R. C., Schepens, C. L., and Swann, D. A., Hyaluronic acid vitreous substitute. A six-year clinical evaluation, *Arch. Ophthalmol.*, 97, 2325–2330 (1979).
30. Buratto, L., Giardini, P., and Bellucci, R., in *Viscoelastics in Ophthalmic Surgery,* Slack Incorporated, Thorofare, NJ, 2000, pp. 313–351.
31. Wobig, M., Dickhut, A., Maier, R., and Vetter, G., Viscosupplementation with hylan G-F 20: a 26-week controlled trial of efficacy and safety in the osteoarthritic knee, *Clin. Ther.*, 20, 410–423 (1998).
32. Brandt, K. D., Smith, G. N., and Simon, L. S., Author reply to Parenti, D., Murray, C.W., Treatment of intraarticular osteoarthritis of the knee with Hylan G-F 20: comment on the article by Brandt et al., *Arthritis Rheum.*, 44, 1473–1476 (2001).

33. Altman, R. D. and Moskowitz, R., Intraarticular sodium hyaluronate (Hyalgan) in the treatment of patients with osteoarthritis of the knee: a randomized clinical trial. Hyalgan Study Group, *J. Rheumatol.*, 25, 2203–2212 (1998).

34. Felson, D. T. and Anderson, J. J., Hyaluronate sodium injections for osteoarthritis: hope, hype, and hard truths, *Arch. Intern. Med.*, 162, 245–247 (2002).

35. Lo, G., LaValley, M., McAlindon, T., and Felson, D., Intra-articular hyaluronic acid in treatment of knee osteoarthritis: a meta-analysis, *JAMA*, 290, 3115–3121 (2003).

36. Puhl, W., Bernau, A., Greiling, H., Kopcke, W., Pforringer, W., and Steck, K. J., Intra-articular sodium hyaluronate in osteoarthritis of the knee: a multicenter, double-blind study, *Osteoarthritis Cartilage*, 1, 233–241 (1993).

37. Lohmander, L. S., Dalen, N., Englund, G., Hamalainen, M., Jensen, E. M., Karlsson, K., Odensten, M., Ryd, L., Sernbo, I., Suomalainen, O., and Tegnander, A., Intra-articular hyaluronan injections in the treatment of osteoarthritis of the knee: a randomized, double blind, placebo controlled multicentre trial, *Ann. Rheum. Dis.*, 55, 424–431 (1996).

38. Brandt, K. D., Block, J. A., Michalski, J. P., Moreland, L. W., Caldwell, J. R., and Lavin, P. T., Efficacy and safety of intraarticular sodium hyaluronate in knee osteoarthritis. ORTHOVISC Study Group, *Clin Orthop.*, 385, 130–143 (2001).

39. Brandt, K. D., Smith, G. N. Jr., and Simon, L. S., Intraarticular injection of hyaluronan as treatment for knee osteoarthritis: what is the evidence?, *Arthritis Rheum.*, 43, 1192–1203 (2000).

40. Lequesne, M., Indices of severity and disease activity for osteoarthritis, *Semin. Arthritis Rheum.*, 20, 48–54 (1991).

41. Lequesne, M. G., The algofunctional indices for hip and knee osteoarthritis, *J. Rheumatol.*, 24, 779–781 (1997).

42. Altman, R., Brandt, K., Hochberg, M., Moskowitz, R., Bellamy, N., Bloch, D. A., Buckwalter, J., Dougados, M., Ehrlich, G., Lequesne, M., Lohmander, S., Murphy, W. A., Jr., Rosario-Jansen, T., Schwartz, B., and Trippel, S., Design and conduct of clinical trials in patients with osteoarthritis: recommendations from a task force of the Osteoarthritis Research Society. Results from a workshop, *Osteoarthritis Cartilage*, 4, 217–243 (1996).

43. Dahlberg, L., Lohmander, L. S., and Ryd, L., Intraarticular injections of hyaluronan in patients with cartilage abnormalities and knee pain: a one-year double-blind, placebo-controlled study, *Arthritis Rheum.*, 37, 521–528 (1994).

44. Creamer, P., Sharif, M., George, E., Meadows, K., Cushnaghan, J., Shinmei, M., and Dieppe, P., Intra-articular hyaluronic acid in osteoarthritis of the knee: an investigation into mechanisms of action, *Osteoarthritis Cartilage*, 2, 133–140 (1994).

45. Sala, S. F. and Miguel, R. E., Intra-articular hyaluronic acid in the treatment of osteoarthritis of the knee: a short-term study, *Eur. J. Rheumatol. Inflamm.*, 15, 33–38 (1995).

46. Carrabba, M., Paresce, E., Angelini, M., Re, K. A., Torchiana, E. E. M., and Perbellini, A., The safety and efficacy of different dose schedules of hyaluronic acid in the treatment of painful osteoarthritis of the knee with joint effusion, *Eur. J. Rheumatol. Inflamm.*, 15, 25–31 (1995).

47. Pham, T., Le Henanff, A., Ravaud, P., Dieppe, P., Paolozzi, L., and Dougados, M., Evaluation of the symptomatic and structural efficacy of a new hyaluronic acid compound, NRD101, in comparison with diacerein and placebo in a 1 year randomised controlled study in symptomatic knee osteoarthritis, *Ann. Rheum. Dis.*, 63, 1611–1617 (2004).

48. Jubb, R., Piva, S., Beinat, L., Dacre, I., and Gishen, P. A one-year, randomised, placebo (saline) controlled clinical trial of 500–730 kDa sodium hyaluronate (Hyalgan) on the radiological change in osteoarthritis of the knee, *Int. J. Clin. Pract.,* 57, 467–474 (2003).

49. Felson, D. T., The verdict favors nonsteroidal antiinflammatory drugs for treatment of osteoarthritis and a plea for more evidence on other treatments, *Arthritis Rheum.,* 44, 1477–1480 (2001).

50. Liang, M. H., Larson, M. G., Cullen, K. E., and Schwartz, J.A., Comparative measurement efficiency and sensitivity of five health status instruments for arthritis research, *Arthritis Rheum.,* 28, 542–547 (1985).

51. Scale, D., Wobig, M., and Wolpert, W., Viscosupplementation of osteoarthritic knees with Hylan: a treatment schedule study, *Curr. Therapeut. Res.,* 55, 220–232 (1994).

52. Lo, G. H., Felson, D. T., and LaValley, M., Intra-articular hyaluronic acid for treatment of osteoarthritis of the knee [Letters], *JAMA,* 291, 1441–1442 (2004).

53. Moskowitz, R. and Altman, R. D., Intra-articular hyaluronic acid for treatment of osteoarthritis of the knee [Letters], *JAMA,* 291, 1440–1441 (2004).

54. Jordan, K. M., Arden, N. K., Doherty, M., Bannwarth, B., Bijlsma, J. W. J., Dieppe, P., Gunther, K., Hauselmann, H., Herrero-Beaumont, G., Kaklamanis, P., Lohmander, S., Leeb, B., Lequesne, M., Mazieres, B., Martin-Mola, E., Pavelka, K., Pendleton, A., Punzi, L., Serni, U., Swoboda, B., Verbruggen, G., Zimmerman-Gorska, I., and Dougados, M., EULAR Recommendations 2003: an evidence based approach to the management of knee osteoarthritis: report of a Task Force of the Standing Committee for International Clinical Studies Including Therapeutic Trials (ESCISIT), *Ann. Rheum. Dis.,* 62, 1145–1155 (2003).

55. Balazs, E. A. and Denlinger, J. L., Viscosupplementation: a new concept in the treatment of osteoarthritis, *J. Rheumatol. Suppl.,* 39, 3–9 (1993).

56. Kawcak, C. E., Frisbie, D. D., Trotter, G. W., McIlwraith, C. W., Gillette, S. M., Powers, B. E., and Walton, R. M., Effects of intravenous administration of sodium hyaluronate on carpal joints in exercising horses after arthroscopic surgery and osteochondral fragmentation, *Am. J. Vet. Res.,* 58, 1132–1140 (1997).

57. Pozo, M. A., Balazs, E. A., and Belmonte, C., Reduction of sensory responses to passive movements of inflamed knee joints by Hylan, *Exp. Brain Res.,* 116, 3–9 (1997).

58. Gomis, A., Pawlak, M., Balazs, E. A., Schmidt, R. F., and Belmonte, C., Effects of different molecular weight elastoviscous hyaluronan solutions on articular nociceptive afferents, *Arthritis Rheum.,* 50, 314–326 (2004).

59. Ghosh, P. and Guidolin, D., Potential mechanism of action of intra-articular hyaluronan therapy in osteoarthritis: are the effects molecular weight dependent? *Semin. Arthritis Rheum.,* 32, 10–37 (2002).

60. Asari, A., Miyauchi, S., Matsuzaka, S., Ito, T., Kominami, E., and Uchiyama, Y., Molecular weight-dependent effects of hyaluronate on the arthritic synovium, *Arch. Histol. Cytol.,* 61, 125–135 (1998).

61. Wobig, M., Bach, G., Beks, P., Dickhut, A., Runzheimer, J., Schwieger, G., Vetter, G., and Balazs, E. A., The role of elastoviscosity in the efficacy of viscosupplementation for osteoarthritis of the knee: a comparison of Hylan G-F 20 and a lower-molecular-weight hyaluronan, *Clin. Ther.,* 21, 1549–1562 (1999).

62. Allard, S. and O'Regan, M., The role of elastoviscosity in the efficacy of viscosupplementation for osteoarthritis of the knee: a comparison of Hylan G-F 20 and a lower-molecular-weight hyaluronan, *Clin. Ther.,* 22, 792–795 (2000).

63. Karlsson, J., Sjogren, L. S., and Lohmander, L. S., Comparison of two hyaluronan drugs and placebo in patients with knee osteoarthritis, a controlled, randomized, double-blind, parallel-design multicentre study, *Rheumatology*, 41, 1240–1248 (2002).
64. Knudson, W., Casey, B., Nishida, Y., Eger, W., Kuettner, K. E., and Knudson, C. B., Hyaluronan oligosaccharides perturb cartilage matrix homeostasis and induce chondrocytic chondrolysis, *Arthritis Rheum.*, 43, 1165–1174 (2000).
65. Lindqvist, U., Tolmachev, V., Kairemo, K., Astrom, G., Jonsson, E., and Lundqvist, H., Elimination of stabilised hyaluronan from the knee joint in healthy men, *Clin. Pharmacokinet.*, 41, 603–613 (2002).
66. Tulandi, T., Introduction — prevention of adhesion formation: the journey continues, *Hum. Reprod. Update*, 7, 545–546 (2001).
67. Adzick, N. S. and Longaker, M. T., Scarless fetal healing. Therapeutic implications, *Ann. Surg.*, 215, 3–7 (1992).
68. Longaker, M. T., Chiu, E. S., Adzick, N. S., Stern, M., Harrison, M. R., and Stern, R., Studies in fetal wound healing. V. A prolonged presence of hyaluronic acid characterizes fetal wound fluid, *Ann. Surg.*, 213, 292–296 (1991).
69. Shimanuki, T., Nishimura, K., Montz, F. J., Nakamura, R. M., and diZerega, G. S., Localized prevention of postsurgical adhesion formation and reformation with oxidized regenerated cellulose, *J. Biomed. Mater. Res.*, 21, 173–185 (1987).
70. Sawada, T., Tsukada, K., Hasegawa, K., Ohashi, Y., Udagawa, Y., and Gomel, V., Cross-linked hyaluronate hydrogel prevents adhesion formation and reformation in mouse uterine horn model, *Hum. Reprod.*, 16, 353–356 (2001).
71. Urman, B., Gomel, V., and Jetha, N., Effect of hyaluronic acid on postoperative intraperitoneal adhesion formation in the rat model, *Fertil. Steril.*, 56, 563–567 (1991).
72. Johns, A., Evidence-based prevention of post-operative adhesions, *Hum. Reprod. Update*, 7, 577–579 (2001).
73. Diamond, M. P., Reduction of adhesions after uterine myomectomy by Seprafilm membrane (HAL-F): a blinded, prospective, randomized, multicenter clinical study. Seprafilm Adhesion Study Group, *Fertil. Steril.*, 66, 904–910 (1996).
74. Beck, D. E., The role of Seprafilm bioresorbable membrane in adhesion prevention, *Eur. J. Surg. Suppl.*, 577, 49–55 (1997).
75. Beck, D. E., Cohen, Z., Fleshman, J. W., Kaufman, H. S., van Goor, H., and Wolff, B. G., A prospective, randomized, multicenter, controlled study of the safety of Seprafilm adhesion barrier in abdominopelvic surgery of the intestine, *Dis. Colon Rectum.*, 46, 1310–1319 (2003).
76. Kudo, F. A., Nishibe, T., Miyazaki, K., Murashita, T., Nishibe, M., and Yasuda, K., Use of bioresorbable membrane to prevent postoperative small bowel obstruction in transabdominal aortic aneurysm surgery, *Surg. Today*, 34, 648–651 (2004).
77. Miller, R. S., Steward, D. L., Tami, T. A., Sillars, M. J., Seiden, A. M., Shete, M., Paskowski, C., and Welge, J., The clinical effects of hyaluronic acid ester nasal dressing (Merogel) on intranasal wound healing after functional endoscopic sinus surgery, *Otolaryngol. Head Neck Surg.*, 128, 862–869 (2003).
78. Catalano, P. J. and Roffman, E. J., Evaluation of middle meatal stenting after minimally invasive sinus techniques (MIST), *Otolaryngol. Head Neck Surg.*, 128, 875–881 (2003).
79. Gaskins, R. E., Scarring in endoscopic ethmoidectomy, *Am. J. Rhinol.*, 8, 271–274 (1994).

80. Jacob, A., Faddis, B. T., and Chole, R. A., MeroGel hyaluronic acid sinonasal implants: osteogenic implications, *Laryngoscope,* 112, 37–42 (2002).
81. Martini, A., Morra, B., Aimoni, C., and Radice, M., Use of a hyaluronan-based biomembrane in the treatment of chronic cholesteatomatous otitis media, *Am. J. Otol.,* 21, 468–473 (2000).
82. Li, G., Feghali, J. G., Dinces, E., McElveen, J., and van de Water, T. R., Evaluation of esterified hyaluronic acid as middle ear-packing material, *Arch. Otolaryngol. Head Neck Surg.,* 127, 534–539 (2001).
83. Stenfors, L. E., Treatment of tympanic membrane perforations with hyaluronan in an open pilot study of unselected patients, *Acta Otolaryngol. Suppl.,* 442, 81–87 (1987).
84. Stenfors, L. E., Repair of traumatically ruptured tympanic membrane using hyaluronan, *Acta Otolaryngol. Suppl.,* 442, 88–91 (1987).
85. Stenfors, L. E., Repair of tympanic membrane perforations using hyaluronic acid: an alternative to myringoplasty, *J. Laryngol. Otol.,* 103, 39–40 (1989).
86. Stenfors, L. E., Berghem, L., Bloom, G. D., Hellstrom, S., and Soderberg, O., Exogenous hyaluronic acid (Healon) accelerates the healing of experimental myringotomies, *Auris Nasus Larynx,* 12, S214–S215 (1985).
87. Diamond, M. P., Reduction of de novo postsurgical adhesions by intraoperative precoating with Sepracoat (HAL-C) solution: a prospective, randomized, blinded, placebo-controlled multicenter study. The Sepracoat Adhesion Study Group, *Fertil. Steril.,* 69, 1067–1074 (1998).
88. Thornton, M. H., Johns, D. B., Campeau, J. D., Hoehler, F., and DiZerega, G. S., Clinical evaluation of 0.5% ferric hyaluronate adhesion prevention gel for the reduction of adhesions following peritoneal cavity surgery: open-label pilot study, *Hum. Reprod.,* 13, 1480–1485 (1998).
89. Johns, D. B., Keyport, G. M., Hoehler, F., and diZerega, G. S., Reduction of postsurgical adhesions with Intergel adhesion prevention solution: a multicenter study of safety and efficacy after conservative gynecologic surgery, *Fertil. Steril.,* 76, 595–604 (2001).
90. The American Fertility Society classifications of adnexal adhesions, distal tubal occlusion, tubal occlusion secondary to tubal ligation, tubal pregnancies, mullerian anomalies, and intrauterine adhesions, *Fertil. Steril.,* 49, 944–955 (1988).
91. Kimmelman, C. P., Edelstein, D. R., and Cheng, H. J., Sepragel sinus (Hylan B) as a postsurgical dressing for endoscopic sinus surgery, *Otolaryngol. Head Neck Surg.,* 125, 603–608 (2001).
92. Haney, A. F., Hesla, J., Hurst, B. S., Kettel, L. M., Murphy, A. A., Rock, J. A., Rowe, G., and Schlaff, W. D., Expanded polytetrafluoroethylene (Gore-Tex Surgical Membrane) is superior to oxidized regenerated cellulose (Interceed TC7+) in preventing adhesions, *Fertil. Steril.,* 63, 1021–1026 (1995).
93. Haney, A. F. and Doty, E., A barrier composed of chemically cross-linked hyaluronic acid (Incert) reduces postoperative adhesion formation, *Fertil. Steril.,* 70, 145–151 (1998).
94. Schlussel, R., Cystoscopic correction of reflux, *Curr. Urol. Rep.,* 5, 127–131 (2004).
95. Puri, P., Chertin, B., Velayudham, M., Dass, L., and Colhoun, E., Treatment of vesicoureteral reflux by endoscopic injection of dextranomer/hyaluronic acid copolymer: preliminary results, *J. Urol.,* 170, 1541–1544 (2003).
96. Chertin, B. and Puri, P., Endoscopic management of vesicoureteral reflux: does it stand the test of time?, *Eur. Urol.* 42, 598–606 (2002).

97. Oswald, J., Riccabona, M., Lusuardi, L., Bartsch, G., and Radmayr, C., Prospective comparison and 1-year follow-up of a single endoscopic subureteral polydimethylsiloxane versus dextranomer/hyaluronic acid copolymer injection for treatment of vesicoureteral reflux in children, *Urology*, 60, 894–897 (2002).

98. Gertzman, A. A. and Sunwoo, M. H. A pilot study evaluating sodium hyaluronate as a carrier for freeze-dried demineralized bone powder, *Cell Tissue Bank.*, 2, 87–94 (2001).

99. Prato, G. P., Rotundo, R., Magnani, C., Soranzo, C., Muzzi, L., and Cairo, F., An autologous cell hyaluronic acid graft technique for gingival augmentation: a case series, *J. Periodontol.*, 74, 262–267 (2003).

100. Schultz, R. H., Wollen, T. H., Greene, N. D., Nathan, D., Brown, K. K., and Mozier, J. O., Remote administration of hyaluronic acid to mammals. *USP 4,808,576* (1989).

Index

A

Abdominal surgeries, *see* postsurgical adhesions
Absolute viscosity, 87–91
Absorbant Intranasal Packing Material (Absorbant Intranasal Splint), 191
Acetylation, 27, 44, 117
Acid hydrolysis, 1, 36–37, 113, 127
Acylation, 27, 28, 30
Adhesion prevention, *see* postsurgical adhesions
Adipic dihydrazide (ADH), 51, *52*, 54, 55, 58
Adverse events (AE), 135, 136
　abscesses, 141
　allergic reaction, 149, 197
　bowel obstruction, 190, 196
　foreign-body reaction, 138, 149
　infections, 143–144, 197
　infertility, 189, 196
　MAUDE safety information, 139
　pseudogout, 137
　SAIRs (severe acute inflammatory reactions), 137–138
　skin testing, 147, 149, 151
Aerosolized HA product, 18
Alcian Blue Assay, 105
Aldehydes, 68, 69
　chromogenic reactions, 31, 34
　peroxidation, 30, 31
　reactions, 45, 46, 59
　reduction, 35, 117
　Ugi reaction, 48, 49, 61
Alginic acid (AA), 61, 62
Alkalines, 32, 69
　deacetylation, 43
　decreased viscosity of HA, 84
　degradation of HA, 34, 35
　HA crosslinking, 56, 58, 59
　phosphotase reactions, 106
Alkylamine bridge, 46, 47
Alkylation, 27, 28
Allograft tissues, 19, 201

American Academy of Ophthalmology (AAO), 99
American College of Rheumatology (ACS), 17
American Fertility Society (AFS), 195
American National Standards Institute (ANSI), 99
Amvisc®; *see also* Amvisc Plus, 15, 16
　aspiration, 160
　coating ability, 162
　development using rabbit eyes, 114
　FDA approval, 125
　indications, 131
　safety, 132, 133, 150
　viscosity, 88
Amvisc Plus, 112, 131, 159, 161, 163
Antitrypsin deficiency, 130
Artz, 136, 183

B

Bacteria; *see also* endotoxins, 115
　Gram-negative, 115, 116
　Gram-positive, 116
　hyaluronidases, 2, 3, 63
　streptococci, 2, 9, 11
　wound infection, 143–144
Bacterial fermentation, 11, 16
Balanced salt solution (BSS), 132, 159, 164, 165, 166
Biocompatibility, 10, 99–102, 125
　with chemical modification, 129
　testing methods, 101, 103
Biological evaluation, *see* biocompatibility
BioLon®, 15, 16
　indications, 131
　osmolality, 112
　safety, 133
Biosynthesis, 11–12, 43, 116
Biotinylated HA, 51–52
Bis-aldehyde, 58
Blue Book Memos, 99, 100
BODIPY-labeled HA, 50